HEDGING DERIVATIVES

ADVANCED SERIES ON STATISTICAL SCIENCE & APPLIED PROBABILITY

Editor: Ole E. Barndorff-Nielsen

Advanced Series on

Statistical Science &

Applied Probability

Vol. 15

HEDGING DERIVATIVES

Thorsten Rheinländer

London School of Economics and Political Science, UK

Jenny Sexton

University of Manchester, UK

 **World Scientific**

NEW JERSEY · LONDON · SINGAPORE · BEIJING · SHANGHAI · HONG KONG · TAIPEI · CHENNAI

Published by

World Scientific Publishing Co. Pte. Ltd.

5 Toh Tuck Link, Singapore 596224

USA office: 27 Warren Street, Suite 401-402, Hackensack, NJ 07601

UK office: 57 Shelton Street, Covent Garden, London WC2H 9HE

Library of Congress Cataloging-in-Publication Data
Rheinländer, Thorsten.
    Hedging derivatives / by Thorsten Rheinländer & Jenny Sexton.
      p. cm. -- (Advanced series on statistical science and applied probability ; v. 15)
    Includes bibliographical references and index.
    ISBN-13: 978-981-4338-79-0 (alk. paper)
    ISBN-10: 981-4338-79-6 (alk. paper)
    1. Hedging (Finance)--Mathematical models. 2. Derivative securities--Valuation--Mathematical
models. I. Sexton, Jenny. II. Title.
    HG6024.A3.R517 2011
    332.64'57--dc22

                                                                            2011004304

British Library Cataloguing-in-Publication Data
A catalogue record for this book is available from the British Library.

Printed in Singapore.

Preface

Hedging of derivatives refers to approximating the payoff of a derivative security. This can be achieved by trading dynamically in the underlying financial asset, or by holding a portfolio of simpler derivatives like plain vanilla options. In a complete market the payoff of all derivative securities can be perfectly replicated using a self-financing dynamic trading strategy. In particular, writing derivatives on complete markets would not increase the range of risk exposures available to investors, which is not realistic. On the other hand, in an incomplete market by definition there exist derivatives which cannot be perfectly replicated and we have to content ourselves with an approximation. Incomplete markets thus introduce an unavoidable additional risk into trading derivatives and one has to aim to minimise this residual exposure.

This book provides an overview of some of the unique and challenging features of incomplete markets as well as the techniques used to derive optimal hedging strategies. A range of criteria can be used to minimise the residual unhedgeable risk. The connections between these approaches and their associated optimal martingale measures is discussed. Two key examples are the mean-variance and utility indifference criteria. Most results are presented in a general framework and then applied to two classes of market models: where the underlying asset is either an exponential Lévy process or a continuous process with stochastic volatility. Some concrete examples are provided using insurance derivatives and markets with basis risk. Dynamic trading is not essential to hedging and sometimes it is preferable to construct a static portfolio of simpler derivatives to approximate a more complicated derivative. Exact replication of exotic derivatives by semi-static hedging requires the underlying asset price process to exhibit certain symmetry properties.

vi   *Hedging Derivatives*

Our presentation proceeds on a medium level of difficulty and assumes a basic understanding of stochastic integration with respect to Brownian motion and the Black & Scholes theory of option pricing. The distinction between local and true martingales is treated in detail since this is related to arbitrage and financial bubble phenomena. However, we do not present all results in full generality but rather aim for transparent proofs. Likewise, we have avoided the need for general semi-martingale characteristics and do not employ too much of the machinery of BMO-martingales because we feel that this would render the exposition too technical.

References for further reading can be found at the end of each chapter. We do not aim for a historical account of the subject matter but have rather included the most comprehensive recent references.

The book provides background material for early stage researchers and could be used for a two-term postgraduate course. Some aspects of the text have been taught within the MSc Risk & Stochastics at the London School of Economics, and we are grateful for many suggestions from the students.

Thorsten Rheinländer & Jenny Sexton
London, January 2011

Contents

Chapter 1

Introduction

A derivative is a financial instrument whose payoff depends on an underlying asset with price process S. The most basic derivatives, also called vanilla options, are calls and puts with payoff $C = (S_T - K)^+$, respectively $P = (K - S_T)^+$. Both strike price K and maturity T are specified in the derivative contract. A forward contract has payoff $F = S_T - K$, so in contrast to the call and put can assume negative values. Since the payoff of these claims depends only on the asset price at maturity, they are referred to as European options.

Calls, puts and futures can be considered in a way to be the atoms of more general European claims. Many common European options like straddles or butterfly spreads can be created as linear combinations of vanilla options. More generally, when the payoff of a European option is a convex function, it can be represented as a portfolio of vanillas if one allows for a continuum of strike prices. This representation allows us without the need for any model specification to approximate the payoff of more complicated options using vanilla options. The hedging approach introduced so far requires trading to take place only at inception. Path dependent options such as barrier options require a more sophisticated method, called semi-static hedging, which allows for trading at an intermediary time. Semi-static hedging can only achieve perfect replication of a derivative when the underlying asset's returns satisfy certain symmetry properties.

An alternative approach is to hedge options by dynamically trading in the underlying asset with the aim of offsetting any losses from the derivative by gains from investing in the underlying asset. Dynamic hedging portfolios should be adjusted whenever the asset price moves, but in practice the portfolio is adjusted at discrete points in time which could be chosen according to some optimality criterion.

1.1   Hedging in complete markets

This section provides a short overview of the mathematical techniques underpinning dynamic hedging in complete markets. Our review is intentionally kept short, the next two chapters contain a more detailed exposition of the concepts and results deployed here.

1.1.1   *Black & Scholes analysis and its limitations*

Classically asset prices are modelled as having log-returns which are normally distributed. In this 'Samuelson model', the discounted asset price process S is a geometric Brownian motion, with constant drift and volatility coefficients, denoted μ and $\sigma > 0$. We work on a probability space $(\Omega, \mathcal{F}, P)$, and formally define S to be the solution to the stochastic differential equation (SDE)

$$dS = \mu S \, dt + \sigma S \, dB$$

where B is a standard Brownian motion. This SDE has the explicit solution

$$S = \exp\left(\sigma B + \left(\mu - \frac{1}{2}\sigma^2\right)t\right),$$

and hence the filtration $\mathbb{F}$ generated by the price process S equals the Brownian filtration. Typically, the price process S will on average have a trend so is not necessarily a martingale. However, S may be transformed into a martingale using a change of measure.

For this we require the stochastic exponential of $-\frac{\mu}{\sigma}B$, which is given by

$$Z = \mathcal{E}\left(-\frac{\mu}{\sigma}B\right) = \exp\left(-\frac{\mu}{\sigma}B - \frac{1}{2}\frac{\mu^2}{\sigma^2}t\right),$$

and satisfies the stochastic integral equation

$$Z = 1 - \frac{\mu}{\sigma}\int Z \, dB.$$

This can be verified by applying the time-dependent Itô formula to $f(t, B_t)$ with $f(t, x) = \exp(-\mu x/\sigma - \mu^2 t/2\sigma^2)$. As Z is a stochastic integral with respect to B, it is a local martingale. In fact, by Novikov's criterion Z is even a martingale. Therefore, Z can act as the density process of a measure change.

Define a new probability measure Q equivalent to P using $dQ/dP = Z_T$. The Q-probability of any set $\Gamma \in \mathcal{F}_T$ is then

$$Q(\Gamma) = E[Z_T \mathbb{I}_\Gamma].$$

A consequence of Girsanov's theorem is that $\widetilde{B} = B + \mu t/\sigma$ is a Q-martingale. Moreover, since the quadratic variation of $\widetilde{B}$ is $[\widetilde{B}]_t = [B]_t = t$, Lévy's characterisation tells us that $\widetilde{B}$ is a Q-Brownian motion.

We associate with any European claim $H = h(S_T)$ a Q-martingale V, called the value process of H, by taking the Q-conditional expectation,

$$V_t = E_Q[H \,|\, \mathcal{F}_t].$$

Since $\mathbb{F}$ is a Brownian filtration, every $(Q, \mathbb{F})$-martingale can be written as a stochastic integral with respect to $\widetilde{B}$. In particular, the Q-dynamics of S are $dS = \sigma S \, d\widetilde{B}$, so

$$V = V_0 + \int \vartheta^H \, dS$$

for some predictable process ϑ^H. Taking Q-expectations of V_T and using that $\int \vartheta^H \, dS$ is a Q-martingale with Q-expectation equal to zero we obtain

$$V_0 = E_Q[H].$$

The fair price (or 'Black-Scholes price') of the claim H is the expectation of H under the martingale measure Q, which is also commonly called the risk-neutral measure. It would be a severe mistake to take the expectation under the statistical (sometimes referred to as 'real-world') measure P unless the drift μ is zero. The integrand ϑ^H can be interpreted as replicating hedging strategy. When H is a vanilla call or put option, the Q-expectation may be explicitly computed in terms of the normal distribution cdf. Furthermore, the replicating strategy can be related to a partial differential equation.

Geometric Brownian motion is a highly tractable but not particularly realistic model for asset prices. For instance, implied volatility can be calculated by inverting the Black-Scholes option pricing formula and using the market prices of vanilla options to derive the volatility in the market. Typically, this implied volatility is not constant. Instead, a smile effect is observed whereby options in or out of the money have a higher implied volatility than options at the money. Skewness effects can also be observed, whereby the implied volatility curve is downward sloping. Moreover, log-returns of equities over short time periods are typically not normally distributed.

1.1.2   *Complete markets*

The general features of the approach to dynamic hedging described above
may be summarised as follows:

(1) An asset price process S is specified such that there exists an equivalent
    martingale measure Q under which S is a (local) Q-martingale. The
    first fundamental theorem of asset pricing states that the existence of
    such a measure is essentially equivalent to the absence of arbitrage
    opportunities. That is, there is no way to make a gain by investing in
    the risky asset S without incurring some risk.
(2) Each claim is associated with a value process V by taking conditional
    expectations with respect to Q. The value process V is a Q-martingale
    by the tower property of conditional expectation. If the filtration gen-
    erated by S satisfies the predictable representation property, V can be
    written as a stochastic integral with respect to S. Thus the claim can
    be perfectly replicated by dynamically investing in S, and the integrand
    in the representation of V is the hedging strategy. When every claim
    is attainable in this sense, the financial market is called complete.
(3) The second fundamental theorem of asset pricing states that an
    arbitrage-free market is complete if and only if there exists a unique
    equivalent martingale measure Q. Hence, the martingale measure used
    in the first step is unique so may be taken as the pricing measure. The
    fair price of each claim is then its expectation under Q.

As we have seen, when the asset price is modelled as geometric Brownian
motion the market is complete. Also, a negative compensated Poisson
process induces a complete market. Further examples of complete markets
include local volatility models where the price process under a martingale
measure Q is given as solution to a SDE of the form $dS_t = \sigma(S_t)\,dB_t$.
Interestingly, some pure jump models based on Azéma martingales are also
complete.

1.2   Hedging in incomplete markets

1.2.1   *Sources of incompleteness*

Empirical studies suggest that real markets exhibit features that require
asset pricing models beyond the paradigm of complete markets. Exponential Lévy models are one of the most popular alternatives and model

log-returns using a stationary process with independent increments like the variance-gamma, normal inverse Gaussian or hyperbolic process. In general, exponential Lévy processes generate incomplete markets since there is uncertainty about both the timing and size of the jumps in the paths of these processes. In fact, the only non-trivial Lévy martingales with the predictable representation property are Brownian motion, which has continuous paths, and the compensated Poisson process that has jump size equal to one.

An even better fit to empirical data can be achieved by stochastic volatility models. The asset price and volatility processes can be driven by multiple sources of noise and hence, in general, these models also lead to incomplete markets. Stochastic volatility models can offer greater modelling flexibility since they display a rich variety of behaviour, including asset price bubbles, which are absent from the Lévy world. Moreover, in continuous stochastic volatility models, if an option can be written such that its price is a convex function of the underlying asset price, then the market can be completed by trading in both underlying asset and the option written on it.

Finally, often there may be constraints on the hedging strategies which can be executed. For example, there may be limits placed on the size of a single position, the composition of portfolios or prohibitions on short selling. Furthermore, in some markets derivatives are written on underlying assets which are not traded, so the only possibility is to hedge using a correlated asset. Even when the model generated by the asset price is complete, trading constraints could prevent the construction of the replicating strategy.

1.2.2 *Calibration*

In incomplete markets, the second fundamental theorem of asset pricing implies that there exist multiple martingale measures. However, in the presence of a liquid market of actively traded vanilla options written on the underlying asset, it is sometimes possible to identify the martingale measure 'chosen by the market' via calibration. Before commencing to calibrate, a parametric family of asset price models must be selected. The modelling takes place in the risk-neutral world, so that each set of parameters corresponds to one particular choice of martingale measure. Theoretical option pricing formulae for vanilla puts and calls can then be derived. The aim is to find a parameter set so that the theoretical option prices are as close as possible to the observed market prices of a set of vanilla options. If a single closest match can be achieved, this procedure specifies uniquely an instantaneous market martingale measure.

In practice, calibration is not at all straightforward as typically the problem is ill-posed. In particular, the problem is not convex and the resulting minimisation problem need not have a unique solution, so a gradient descent search might get trapped in a local minimum. Therefore, it is advisable to include some a priori information like the result of yesterday's calibration. This can be achieved by adding a relative entropy term which penalises pricing measures which are too far away from recent measures. The results of calibration suggest that in the case of exponential Lévy models, the density of the Lévy measure is heavily skewed, and extreme jump sizes are significantly more likely in the risk-neutral world.

1.2.3 *Mean-variance hedging*

Real markets are genuinely incomplete as even though it might be possible to fix a martingale measure for the purpose of pricing via calibration, it will still not be possible to replicate every option payoff. In the mean-variance approach, the remaining risk is minimised according to a quadratic criterion. At first sight, this seems not to be a good idea, since in finance it is natural to be more concerned by the magnitude of losses as opposed to gains and this aversion to losses cannot be captured by a symmetric criterion. However, when this mean-variance criterion is formulated under a risk-neutral measure, it turns out that this is, in fact, quite a viable approach to hedging. Beginning under the statistical measure, mean-variance hedging consists of two steps: firstly, one changes to the risk-neutral measure, which introduces a high degree of asymmetry as discussed in the section about calibration; then one follows a quadratic approach under the martingale measure. Moreover, this approach is asymptotically linked to a utility-indifference criterion under the statistical measure.

In mean-variance hedging, orthogonal projection is used to decompose claims into an attainable part which can be fully hedged, and a non-attainable part which constitutes the intrinsic risk of the claim. It is the variance of this residual risk which is most important to a risk manager when making a decision whether to offer a financial product to a client or not. In fact, the results of simulations undertaken by Hubalek *et al.* (2006) and Kallsen & Pauwels (2010b) for both exponential Lévy and stochastic volatility models show that the mean-variance price and the optimal hedging strategy are fairly robust with respect to model risk, whereas the residual risk is very sensitive to the choice of model. In the extreme case of complete market models, this intrinsic risk is zero which is extremely unrealistic.

1.2.4   *Utility indifference pricing and hedging*

A utility function quantifies an agent's preferences about risk exposure. Agents with distinct risk preferences solve separate portfolio allocation problems to determine how to allocate their capital between risky underlying assets and a bank account. The use of utility functions to value random endowments originates in actuarial mathematics as the 'premium principle of equivalent utility'. The concept is to charge a price which ensures that the agent's utility is unchanged by having bought (or sold) a contingent claim. Hence, to calculate a utility indifference price and hedge, it is necessary to solve a pair of portfolio allocation problems. We focus on the family of exponential utility functions because the resulting prices have desirable qualitiative properties, see Gerber (1979).

In a complete market setting, this approach leads to the standard risk-neutral valuation and hedge. However, when the market is incomplete the utility indifference price and hedge depend on the choice of utility function. The risk aversion parameter reflects the agent's tolerance of the unhedgeable part of random endowments. Consequently, agents who are more risk averse will demand a higher price for a claim that cannot be perfectly replicated as compensation for accepting the inaccessible risk. In the limit, as an agent becomes ever more risk averse, he tends towards requiring the super-replication price for the claim.

The agent's portfolio optimisation problem can be reformulated in terms of minimising an entropic criterion over the family of equivalent martingale measures. In the absence of a claim, this dual problem is solved by the minimal entropy martingale measure. The case with a random endowment can be reduced to this case by a measure transform. As the agent's risk aversion decreases and she becomes risk neutral, a linear pricing rule under the entropy measure is recovered. The utility-indifference hedging strategy for a risk neutral agent corresponds to the mean-variance hedging strategy under the entropy measure.

An alternative approach to solving a portfolio allocation problem hinges on the observation that the optimal portfolio's value function is a martingale. The martingale optimality principle can be used to reduce a portfolio optimisation problem to the task of finding the solution to a certain quadratic backward stochastic differential equation (BSDE). This backward formulation is required as the distributional properties of the derivative price are unknown prior to maturity. This method can be adapted to incorporate trading constraints such as basis risk or stochastic volatility. In the absence of trading constraints, these two approaches to utility indifference pricing and hedging are intimately linked and the minimal entropy measure can be characterised in terms of the solution to a quadratic BSDE.

1.2.5 *Exotic options*

In principle, the dynamic approach to pricing and hedging described so far applies equally well to exotic path-dependent options as well as to European options. However, the robustness of exotic option prices and the corresponding hedges to model risk is questionable. In particular, although it may be possible to achieve semi-explicit mean-variance hedges for certain path-dependent insurance derivatives (see Section 6.4) it would be both difficult and time consuming to semi-explicitly calculate the orthogonal projections of each specific exotic option.

Even within complete markets, there is a plethora of articles on dynamic approaches to pricing and hedging of specific exotics such as Asian options, barrier options and lookback options. Asian options are a convenient choice in volatile markets as the payoff depends on an average of past asset prices. Lookback and barrier options depend on the running maximum (or minimum) of the asset price process, the former are written on the minimum stock price up-to-date while the second feature payoffs which are contingent on whether certain price levels have been surpassed or not.

In this book, we follow a different approach consisting of firstly noting that the number of distinct exotic options which require replication can be reduced using duality relationships, and then semi-statically hedging path dependent options. A natural ordering of claims emerges whereby more complex claims can be hedged using more basic assets. Firstly some exotics like barrier options can be replicated using a semi-static portfolio of general European options. In turn, the payoff of these European options can be approximated using vanilla options. Finally, the static portfolio of vanilla options can be dynamically hedged.

The most basic relation between option payoffs is the put-call parity which states that the price of an American call is the sum of the prices of a forward contract and a European put, all with the same strike and maturity. When the price process is a martingale, as opposed to a strict super-martingale, in the risk-neutral world the American call can be replaced by its European counterpart.

Duality relations arise when the price process is taken as a numeraire which leads to the notion of the dual measure and process. The simplest example is the call-put duality which allows to express the price of a call in terms of a put written on the dual process and with inverse strike. In the risk neutral world, certain symmetries in asset returns lead to self-dual asset prices. Self-duality is instrumental in the semi-static hedging of barrier options.

1.2.6 *Optimal martingale measures*

Recently there is increasing activity in new financial markets where a mature market for exchange traded derivatives has yet to emerge and therefore accurate calibration is hard to achieve. Examples include the securitisation of mortality risk as well as the markets for weather derivatives, energy and carbon. In these circumstances it is not feasible to rely on calibration and it is necessary to choose a pricing measure which relates to an optimisation procedure. The most suitable choice of optimisation criterion may depend on the type of claim to be hedged or on personal risk tolerances.

Fundamentally, the choice of an optimal martingale measure should go hand in hand with a corresponding hedging concept. Local hedging concepts are best suited to derivatives which are contingent on events that happen at intermediary random times. Examples of such derivatives include credit, insurance and American-type options. In all these cases, the density of the corresponding optimal martingale measure should be stable with respect to stopping. A martingale measure with this property is the Esscher measure. In contrast, in the case of European claims it is also appropriate to use a global hedging concept such as asymptotic utility-indifference hedging or mean-variance hedging. For example, an asymptotic utility-indifference approach based on the exponential utility function would lead us to choose the minimal entropy martingale measure.

1.3 Notes and further reading

In this section we collect references about important practical topics whose mathematical treatment nevertheless lies outside the scope of this book.

In practice, one cannot continuously readjust a dynamic hedging portfolio, but is confined to trade at discrete time points. Consequently, even in a complete market one can never execute the ideal replicating hedging strategy and is always in an incomplete market setting. One approach is then to minimise the hedging error coming from the discretisation by a quadratic criterion formulated under a martingale measure. This is discussed in Geiss & Geiss (2006) and Geiss & Hujo (2007).

Sensitivities of option prices with respect to model or contractual parameters like the volatility or the time to maturity are commonly referred to as the Greeks. These can be efficiently simulated by Monte Carlo methods based on Malliavin derivatives. See Benth *et al.* (2010) for computations in a stochastic volatility setting and further references.

Calibration of model parameters to observed option prices is treated in Cont & Tankov (2004), (2006). The existence of multiple possible solutions to the calibration problem is related to model uncertainty, see Cont (2006).

Chapter 2

Stochastic Calculus

This chapter serves as reference for results from general stochastic calculus needed throughout this book. The results are often presented without proof. Readers requiring a more detailed account are referred to relevant sections of the monograph by Protter (2005). Results have been selected due to their importance in studying hedging of derivatives. A detailed discussion of the Kunita-Watanabe decomposition is included as this is crucial for mean-variance hedging. Moreover, we provide a general result about the structure of martingale measures.

2.1 Filtrations and martingales

Let us outline some preliminaries that will be assumed to hold throughout the book unless explicitly indicated otherwise. We work on a filtered probability space $(\Omega, \mathcal{F}, \mathbb{F}, P)$. Technically, a **filtration** $\mathbb{F} = (\mathcal{F}_t)$ is an increasing family of σ-algebras $\mathcal{F}_t$ indexed by time, such that $\mathcal{F}_s \subset \mathcal{F}_t$ for $s < t$. A filtration captures the flow of information: the more time elapses, the more knowledge is accumulated. We fix a finite time horizon T and assume $\mathcal{F} = \mathcal{F}_T$; all stochastic processes we will consider are defined on the time interval $[0, T]$. We also assume that $\mathcal{F}_0$ contains only sets of probability zero or one, and that $\mathbb{F}$ fulfills the usual conditions of right-continuity and completeness.

A stochastic process $M = (M_t)$ that is adapted to $\mathbb{F}$ is referred to as a **martingale** with respect to $\mathbb{F}$ and the probability measure P if it fulfills the martingale property:

$$E\left[M_t \,|\, \mathcal{F}_s\right] = M_s, \qquad s \leq t.$$

For this conditional expectation to be well-defined it is necessary to assume that for each t, the random variable M_t is integrable. If this equality is replaced by a '$\geq$' (resp. '$\leq$') we refer to M as a **sub-martingale** (resp. **super-martingale**). (In-)equalities, such as these, between random variables should be understood in the P-almost sure sense.

Example 2.1. Associated martingale. We can associate to each integrable random variable X a martingale by setting:

$$M_t = E\left[X | \mathcal{F}_t\right].$$

The martingale property follows from the tower property of conditional expectation because for all $s \leq t$ we have:

$$\begin{aligned} E\left[M_t | \mathcal{F}_s\right] &= E\left[E\left[X | \mathcal{F}_t\right] | \mathcal{F}_s\right] \\ &= E\left[X | \mathcal{F}_s\right] \\ &= M_s. \end{aligned}$$

Well-known examples of martingales include Brownian motion and the compensated Poisson process. These martingales can be defined with respect to the filtrations they generate, however, these 'natural filtrations' are typically not right-continuous. An **augmented filtration** fulfilling the usual conditions can be constructed by adding all the P-null sets to each $\mathcal{F}_t$. Under the assumption that $\mathbb{F}$ fulfills the usual conditions, it is possible to choose a **cadlag** (i.e. right-continuous with left-limits) version of any sub-martingale M, and we always work with this version without mentioning it explicitly. The **filtration** $\mathbb{F}^X$ **generated** by a stochastic process X is the smallest augmented filtration $\mathbb{F}$ such that X is $\mathbb{F}$-adapted.

The reference filtration and probability measure are essential to the definition of a martingale. For instance, if we enlarge our filtration, there is a priori no reason for the martingale property to hold with respect to the larger filtration. Moreover, the pure set theoretic inclusion $\mathbb{F} \subset \mathbb{G}$ does not reveal much about the relation between the two filtrations. The most important relationship between filtrations, for our purposes, is captured in the following definition:

Definition 2.2. Let $\mathbb{F} \subset \mathbb{G}$ be two filtrations. The smaller filtration $\mathbb{F}$ is said to be **immersed** into the larger filtration $\mathbb{G}$ if every $\mathbb{F}$-martingale is also a $\mathbb{G}$-martingale.

Typically, enlarging a filtration such that the immersion property holds can be considered as being relatively 'harmless' since the structure of the model is not significantly altered. Instead of using the notion of immersion, it is sometimes said that 'Hypothesis (H)' holds but this is somewhat non-descriptive. Secondly, a P-martingale need not be a martingale under an equivalent probability measure Q. This issue will be discussed in Section 2.4.

Definition 2.3. A random variable τ with values in $[0, T]$ is called a **stopping time** if $\{\tau \leq t\} \in \mathcal{F}_t$ for all $t > 0$. The **stopping time σ-algebra** $\mathcal{F}_\tau$ consists of all sets of the form

$$\{\Gamma \in \mathcal{F} \mid \Gamma \cap \{\tau \leq t\} \in \mathcal{F}_t, \text{ all } t \geq 0\}.$$

It is of interest to note that, for an adapted, cadlag process the last exit time from a closed set is not necessarily a stopping time, while the hitting time of an open set is a stopping time. The next result tells us that the martingale property extends to bounded stopping times.

Theorem 2.4. *Optional sampling theorem.* *Let M be a martingale, and let $\sigma \leq \tau$ be two bounded stopping times. Then*

$$E\left[M_\tau \mid \mathcal{F}_\sigma\right] = M_\sigma.$$

We now turn to a discussion of *local martingales*. Local martingales can have very different properties than martingales so can be responsible for pathological behaviour such as arbitrage or stock market bubbles. Subsequently, being able to distinguish between martingales and local martingales is indispensable for a rigorous discussion of financial markets.

Definition 2.5. An adapted stochastic process M is called a **local martingale** if there exists a sequence (T_n) of stopping times, $T_n \uparrow T$, such that M^{T_n} is a martingale for all n. Here M^{T_n} denotes the stopped process given by $M_t^{T_n} = M_{t \wedge T_n}$. The sequence (T_n) is called a **localising sequence** for M. A property of M is said to hold **locally** if it holds for M^{T_n}, each n.

As the local martingale property is somewhat abstract, it is important to have criteria that help us to identify when local martingales are super-martingales.

Proposition 2.6. *Bounded local martingales.* *A local martingale M which is uniformly bounded from below is a super-martingale.*

Proof.   Let (T_n) be a localising sequence for M. Then, for $s \leq t$,

$$E\left[M_t\,|\,\mathcal{F}_s\right] = E\left[\lim_{n\to\infty} M_{t\wedge T_n}\,\middle|\,\mathcal{F}_s\right]$$

$$\leq \lim_{n\to\infty} E\left[M_{t\wedge T_n}\,|\,\mathcal{F}_s\right]$$

$$= \lim_{n\to\infty} M_{s\wedge T_n} = M_s.$$

The first equality uses that (T_n) is a localising sequence. The assumption that M is bounded from below allows expectation and limit to be interchanged by applying the conditional form of Fatou's lemma. The second equality uses that the definition of a local martingale implies the stopped process M^{T_n} is a martingale and the last equality follows since (T_n) is a localising sequence. □

Definition 2.7. The **predictable** σ-**algebra** $\mathcal{P}$ on $\mathbb{R}_+ \times \Omega$ is the σ-algebra generated by the adapted, left-continuous processes. A $\mathcal{P}$-measurable process is called **predictable**.

(A) Brownian motion is predictable since it has continuous paths;
(B) The Poisson process is not predictable.

Here (B) follows because the paths of a Poisson process are right-continuous and jumps occur at totally inaccessible (see Protter (2005), Section III.2, for this concept) stopping times.

   A process is of **finite variation** (FV) if and only if it can be written as difference of two increasing processes. Hence $\int \mu\,dt$, where μ is any process such that the integral exists, is of finite variation since:

$$\int \mu\,dt = \int \mu^+\,dt - \int \mu^-\,dt.$$

The following result illustrates that the only non-trivial continuous martingales are of infinite variation.

Proposition 2.8. *Predictable martingales.* *A local martingale which is predictable and of finite variation is constant.*

2.2   Semi-martingales and stochastic integrals

Definition 2.9. A **semi-martingale** X is an adapted stochastic process which can be written as

$$X = X_0 + M + A \tag{2.1}$$

where M is a local martingale and A a process of finite variation. When $M_0 = A_0 = 0$, (2.1) is called a semi-martingale decomposition of X.

In general this decomposition is not unique. For example, consider the compensated Poisson process $N - \lambda t$, where N is a Poisson process with intensity λ. This process is of finite variation and also a martingale, hence it is a matter of discretion as to whether M or A is identically zero.

Definition 2.10. A **special semi-martingale** is a semi-martingale X with decomposition $X = X_0 + M + A$ such that the finite variation part A is *predictable*. This **canonical decomposition** is unique.

Uniqueness is a simple consequence of Proposition 2.8. Returning to our example of the compensated Poisson process, it is now clear that, since $N - \lambda t$ is not predictable, it forms the martingale part of the canonical decomposition and the finite variation part is identically zero.

Definition 2.11. Let X be a semi-martingale and (Π_n) a sequence of random partitions. Each random partition is a finite sequence of finite stopping times $0 = T_0^n \leq T_1^n \leq ... T_{k_n}^n = T$. The grid size $\|\Pi_n\| = \sup_i |T_{i+1}^n - T_i^n|$ of the sequence (Π_n) is assumed to converge a.s. to zero. The **quadratic variation process** $[X]$ is the increasing, adapted process defined by

$$[X] = \lim_n \sum_i \left(X^{T_{i+1}^n} - X^{T_i^n} \right)^2 \quad \text{in } ucp.$$

The notion of convergence for stochastic processes we use here is *ucp* (i.e. uniformly on compacts in probability). A sequence (X^n) of processes converges to a process X in this sense if $\sup_{0 \leq t \leq T} |X_t^n - X_t|$ converges to 0 in probability.

The **quadratic co-variation** or **square bracket process** $[X, Y]$, for two semi-martingales X and Y, can now be defined by polarisation: $[X, Y] = \frac{1}{2}([X + Y] - [X] - [Y])$. The quadratic co-variation process is a bilinear form which is invariant under equivalent probability measure changes. This follows since a sequence converges in probability if and only if one can extract from every subsequence an almost surely convergent subsequence, and the null sets are the same for equivalent measures. Examples of the quadratic variation of some common processes include:

(A) Brownian motion B: $[B]_t = t$;
(B) Pure jump process X: $[X]_t = \sum_{s \leq t} (\Delta X_s)^2$;
(C) Poisson process N: $[Nd] = N$ since the jump sizes equal one;

(D) Continuous processes X of finite variation: $[X] = 0$.

Definition 2.12. A martingale M is called **square-integrable** (in short: L^2-**martingale**) if $E[[M]_T] < \infty$. A semi-martingale X is called **square-integrable** if it is special with canonical decomposition $X = M + A$ where M is an L^2-martingale and A has square-integrable variation.

Theorem 2.13. *Conditional quadratic variation.* *Let X, Y be two locally square-integrable semi-martingales. Then there exists a unique predictable process $\langle X, Y \rangle$, called the **angle bracket process**, such that $[X, Y] - \langle X, Y \rangle$ is a local martingale. We set $\langle X \rangle := \langle X, X \rangle$ for notational convenience. More generally, $\langle X, Y \rangle$ exists if $[X, Y]$ is of locally integrable variation.*

Observe that since the martingale property is not preserved by a change of measure, the angle bracket process is *not* invariant under measure changes and may even not exist under an equivalent measure. Despite this, it inherits many properties from the square bracket: $\langle X \rangle$ is increasing, $\langle X, Y \rangle$ is a bilinear form, and the angle bracket even coincides with the square bracket when both X and Y are continuous.

(A) Brownian motion: $\langle B \rangle_t = [B]_t = t$;
(B) Poisson process N with intensity λ: $\langle N \rangle_t = \lambda t$.

This second example is a bit tricky, since the equality is not due to the fact that $N - \lambda t$ is a martingale; but rather that $[N] - \lambda t$ is a martingale because $[N] = N$.

Lemma 2.14. *Let X be a locally square-integrable semi-martingale. Then*

$$E[[X]_T] = E[\langle X \rangle_T].$$

Proof. By definition of the angle bracket, $[X] - \langle X \rangle$ is a local martingale. Hence we have for a localising sequence (T_n) that for all n

$$E[[X]_T^{T_n}] = E[\langle X \rangle_T^{T_n}].$$

As both $[X]$ and $\langle X \rangle$ are increasing processes, the result follows by monotone convergence.   □

The following theorem provides conditions which are able to ensure that local martingales are in fact true martingales.

Theorem 2.15. *When are local martingales true martingales?* *A local martingale M is a martingale provided that either:*

(i) $E[\sup_{t \leq T} |M_t|] < \infty$ *or*

(ii) $E[[M]_T] < \infty$, *and in this case we have* $E[M_T^2] = E[[M]_T]$.

Proof. We only prove (i), for (ii) see Protter (2005), Corollary 3 to Theorem II.27. Let (T_n) be a localising sequence for M. By definition, the stopped local martingale M^{T_n} is a martingale, such that for $s \leq t$

$$E[M_{t \wedge T_n} | \mathcal{F}_s] = M_{s \wedge T_n}.$$

Moreover, observe that $\sup_t |M_t| \geq M_{t \wedge T_n}$ for all n, and that we can utilise a dominated convergence argument because by assumption $\sup_t |M_t|$ is integrable:

$$\begin{aligned} E[M_t | \mathcal{F}_s] &= E\left[\lim_{n \to \infty} M_{t \wedge T_n} \,\Big|\, \mathcal{F}_s\right] \\ &= \lim_{n \to \infty} E[M_{t \wedge T_n} | \mathcal{F}_s] \\ &= \lim_{n \to \infty} M_{s \wedge T_n} = M_s. \end{aligned}$$

$\square$

The two conditions in Theorem 2.15 are linked by the Burkholder-Davis-Gundy (BDG) inequalities which we state for the L^p-case.

Proposition 2.16. *BDG inequalities.* *Let M be a local martingale, and $p \geq 1$. Then there exist universal constants c_p and C_p such that*

$$c_p E\left[[M]_T^{p/2}\right] \leq E\left[\sup_{0 \leq t \leq T} |M_t|^p\right] \leq C_p E\left[[M]_T^{p/2}\right].$$

Proposition 2.17. *Doob's inequality.* *Let M be a martingale, $p > 1$. Then*

$$E\left[\sup_{0 \leq t \leq T} |M_t|^p\right] \leq \left(\frac{p}{p-1}\right)^p E[|M_T|^p].$$

At first glance, when applying Theorem 2.15, it might be tempting to attempt to validate condition (i) by estimating the expectation of $\sup_{t \leq T} |M_t|^2$ via Doob's inequality. However, in order to apply Doob's inequality one has to know in advance that M is a martingale! Instead it is necessary to check integrability of quantities related to the whole path of the process, using either the running-maximum or the quadratic variation process $[M]$.

Turning now to stochastic integration, to define the integral with respect to semi-martingale integrators is beyond the scope of this book. Instead we focus on recalling some main properties of stochastic integrals that will be used frequently in the following chapters. Let X be a semi-martingale. In the case when $\int \vartheta\, dX$ is well-defined, we call ϑ **integrable** with respect to X, see Protter (2005), Section IV.2, for details. The set of X-integrable processes, which we will not characterise here, is denoted by $L(X)$. It is important to note that the only reasonable integrands are *predictable* processes. To this end, we often take the left-continuous version ϑ_- instead of ϑ itself to guarantee the resulting stochastic integral is well-defined. The following result is commonly used: if X is locally square-integrable, and ϑ is a predictable process such that

$$P\left(\int_0^T \vartheta_t^2\, d\,[X]_t < \infty\right) = 1,$$

then $\vartheta \in L(X)$.

Theorem 2.18. *Approximation of a stochastic integral.* *Let X be a semi-martingale and ϑ an adapted, cadlag process. Consider a sequence of random partitions (Π_n) with grid size converging to zero. Then*

$$\int \vartheta_-\, dX = \lim_n \sum_i \vartheta_{T_i^n} \left(X^{T_{i+1}^n} - X^{T_i^n}\right) \quad \text{in ucp.}$$

Theorem 2.19. *Quadratic co-variation of stochastic integrals.* *Let X, Y be two semi-martingales, and $\vartheta \in L(X)$, $\eta \in L(Y)$. Then*

$$\left[\int \vartheta\, dX, \int \eta\, dY\right] = \int \vartheta\eta\, d\,[X, Y].$$

In particular, the quadratic variation of the stochastic integral process $\int \vartheta\, dX$ is given by

$$\left[\int \vartheta\, dX\right] = \int \vartheta^2\, d\,[X].$$

Theorem 2.20. *Yoeurp's lemma.* *Let X be a semi-martingale and Y a process of finite variation.*

(i) *If X is a local martingale and Y is predictable, then $[X, Y] = \int \Delta Y\, dX$ is a local martingale.*

(ii) *If either Y or X is continuous, then $[X, Y] = 0$.*

Proof.   Jacod & Shiryaev (2003), Proposition I.4.49.   □

We are now in a position to recall the fundamental rules of calculus for stochastic integrals, and shall start with the product rule.

Theorem 2.21. *Integration by parts.* *Let* X, Y *be two semi-martingales. Then* XY *is a semi-martingale and*

$$XY - X_0Y_0 = \int X_- \, dY + \int Y_- \, dX + [X, Y].$$

The following result is a generalisation of the chain rule.

Theorem 2.22. *Itô's formula.* *Let* X *be a semi-martingale, and* $f :$ $\mathbb{R}_+ \times \mathbb{R} \to \mathbb{R}$ *a* $C^{1,2}$-*function. Then* $f(t, X_t)$ *is a semi-martingale, and we have, with* $[X^c]_t = [X]_t - \sum_{s \leq t} (\Delta X_s)^2$,

$$f(t, X_t) - f(0, X_0) = \int_0^t \frac{\partial}{\partial t} f(s, X_s) \, ds + \int_0^t f'(s, X_{s-}) \, dX_s$$

$$+ \frac{1}{2} \int_0^t f''(s, X_s) \, d[X^c]_s$$

$$+ \sum_{0 < s \leq t} \{ f(s, X_s) - f(s, X_{s-}) - f'(s, X_{s-}) \Delta X_s \}.$$

The next theorem illustrates a useful connection between the space of L^2-martingales and L^2-random variables.

Theorem 2.23. *Itô-isometry.* *For a local martingale* M *and a predictable process* ϑ *with*

$$E \left[\int_0^T \vartheta_t^2 \, d[M]_t \right] < \infty,$$

we have that $\int \vartheta \, dM$ *is a square-integrable martingale, and*

$$E \left[\left(\int_0^T \vartheta_t \, dM_t \right)^2 \right] = E \left[\int_0^T \vartheta_t^2 \, d[M]_t \right].$$

Proof. This follows immediately from Theorem 2.15(*ii*), since the quadratic variation of a stochastic integral is given by the formula in Theorem 2.19. $\square$

As indicated by this result, stochastic integration with respect to a martingale integrator to some extent preserves the martingale property. In most cases, all one can say initially is that the stochastic integral is a local martingale and some additional effort is required to verify whether it is a true martingale.

Theorem 2.24. *Martingale property of stochastic integrals. For a local martingale integrator M, and a locally bounded predictable ϑ, the stochastic integral process $\int \vartheta \, dM$ is a local martingale. The same conclusion holds in case ϑ is adapted and left-continuous with right-limits.*

Proof.   See Protter (2005), Theorems IV.29 and III.33.   □

While this result should be considered to be one of the most important results in the theory of stochastic integration, a similar result does not hold in full generality. This has led to the introduction of the concept of a σ-martingale, see Protter (2005), Section IV.9.

We will sometimes need the stochastic integral with respect to a vectorial continuous semi-martingale. While the integral with respect to a vectorial continuous FV process can be defined as sum over the component-wise one-dimensional integrals, this is in general not sufficient for a vectorial continuous martingale integrator $M = (M^1, \ldots, M^d)'$ (here "$'$" denotes transposition). To see this, just take $d = 2$, $M^2 = -M^1$, and consider a predictable process ϑ such that $\vartheta \notin L(M^1)$. Then the component-wise integrals are not defined, but by all means the integral $\int \vartheta' \, dM$ should be equal to zero. However, if the processes M^i are pairwise orthogonal, that is $[M^i, M^j] = 0$ for $i \neq j$, then the two different concepts coincide, and we write

$$\int \vartheta' \, dM := \sum_{i=1}^{d} \int \vartheta^i \, dM^i.$$

The space $L^2(M)$ is then defined to be the set of all predictable d-dimensional processes ϑ such that

$$E\left[\int_0^T \|\vartheta_t\|^2 \, dM_t\right] < \infty,$$

where $\|.\|$ denotes the Euclidean norm.

2.3   Kunita-Watanabe decomposition

Given a martingale M adapted to the filtration $\mathbb{F}$, we study an orthogonal decomposition of square-integrable $\mathbb{F}$-martingales by projecting them onto a subspace of stochastic integrals with respect to M. The resulting Kunita-Watanabe decomposition is essential to our discussion of mean-variance hedging. This section relies on basic facts about Hilbert spaces which can be found in functional analysis textbooks, such as Conway (1990).

Definition 2.25. The space $\mathcal{M}^2$ of L^2-martingales is the Hilbert space of all martingales M such that $E[[M]_T] < \infty$ (or alternatively, by the BDG inequality, $E[\sup_{t \leq T} M_t^2] < \infty$), equipped with the scalar product

$$(M, N) := E[M_T N_T].$$

$\mathcal{M}^2$ is complete since every L^2-martingale M can be identified isometrically with its final value $M_T \in L^2(P)$. Two martingales M, $N \in \mathcal{M}^2$ are called **orthogonal** if $E[M_T N_T] = 0$. They are called **strongly orthogonal** if MN is a martingale. $\mathcal{M}_0^2$ consists of all $M \in \mathcal{M}^2$ such that $M_0 = 0$, and $\mathcal{M}_{loc}^2$ denotes the space of all processes which are locally in $\mathcal{M}^2$.

Proposition 2.26. *Kunita-Watanabe inequality.* *Let X, Y be two semi-martingales, and ϑ, ψ be two measurable processes. Then*

$$\int_0^T |\vartheta_s| \, |\psi_s| \, |d[X, Y]_s| \leq \left(\int_0^T \vartheta_s^2 \, d[X]_s \right)^{\frac{1}{2}} \left(\int_0^T \psi_s^2 \, d[Y]_s \right)^{\frac{1}{2}}. \quad (2.2)$$

Here $|[X, Y]|$ denotes the total variation process associated with the finite variation process $[X, Y]$ (see Protter (2005), Section I.7). If in addition both X and Y are locally square-integrable, then we also have

$$\int_0^T |\vartheta_s| \, |\psi_s| \, |d\langle X, Y \rangle_s| \leq \left(\int_0^T \vartheta_s^2 \, d\langle X \rangle_s \right)^{\frac{1}{2}} \left(\int_0^T \psi_s^2 \, d\langle Y \rangle_s \right)^{\frac{1}{2}}. \quad (2.3)$$

Proposition 2.27. *Characterisation of strong orthogonality.* *M, $N \in \mathcal{M}^2$ are strongly orthogonal if and only if the following three assertions are true:*

(i) $[M, N]$ is a martingale.
(ii) $E[M_\tau N_\tau] = M_0 N_0$ for every stopping time $\tau \leq T$.
(iii) $\langle M, N \rangle = 0$.

Proof. (i) In the case that MN is a martingale, it follows by integration by parts that $[M, N]$ is a local martingale. By the Kunita-Watanabe inequality (take $\vartheta = \psi \equiv 1$) and the fact that M, $N \in \mathcal{M}^2$, we get by applying Cauchy-Schwarz to the r.h.s. of (2.2) that $\sup |[M, N]|$ is integrable, hence $[M, N]$ is a martingale. On the other hand, if $[M, N]$ is a martingale, then so is MN by Theorem 2.15 and polarisation. Necessity in (ii) follows from the optional sampling theorem. As for sufficiency, consider for $t \in [0, T]$ and $\Gamma \in \mathcal{F}_t$ the stopping time τ_Γ which equals t if $\omega \in \Gamma$ and T if $\omega \notin \Gamma$. Then

$$E\left[M_t N_t \mathbb{I}_\Gamma\right] + E\left[M_T N_T \mathbb{I}_{\Gamma^c}\right] = E\left[M_{\tau_\Gamma} N_{\tau_\Gamma}\right] = M_0 N_0$$

by assumption, from which, together with $E\left[M_T N_T\right] = M_0 N_0$, it follows that $E\left[M_t N_t \mathbb{I}_\Gamma\right] = E\left[M_T N_T \mathbb{I}_\Gamma\right]$, hence MN is a martingale. The last assertion follows from (i) and the definition of the angle bracket process. This is well-defined since $M, N \in \mathcal{M}^2$ implies, by the Kunita-Watanabe inequality, that $[M, N]$ is of locally integrable variation. □

Definition 2.28. Let $M \in \mathcal{M}_{loc}^2$. The space $L^2(M)$ consists of all predictable processes ϑ such that

$$\|\vartheta\|_{L^2(M)} = E\left[\int_0^T \vartheta_t^2 \, d[M]_t\right]^{\frac{1}{2}} < \infty.$$

The space of all stochastic integral processes $\int \vartheta \, dM$ such that $\vartheta \in L^2(M)$ is denoted by $\mathcal{S}(M)$ and referred to as the **stable subspace** generated by M.

Proposition 2.29. Stable subspace of stochastic integrals. $\mathcal{S}(M)$ *is a closed subspace of* $\mathcal{M}^2$. *If* $\int \vartheta \, dM \in \mathcal{S}(M)$, *and* τ *is a stopping time, then the stopped martingale* $\left(\int \vartheta \, dM\right)^\tau$ *is in* $\mathcal{S}(M)$ *as well.*

Proof. $L^2(M)$ can be identified with the L^2-space over the product measure $dP \otimes d[M]$. As such it is a Hilbert space with respect to the usual L^2-scalar product which induces the norm $\|.\|_{L^2(M)}$. On the other hand, $\mathcal{S}(M)$ is a subspace of $\mathcal{M}^2$ since $\left[\int \vartheta \, dM\right]_T = \int_0^T \vartheta_t^2 \, d[M]_t$ is integrable for $\vartheta \in L^2(M)$. It follows by the Itô-isometry that

$$\left\|\int \vartheta \, dM\right\|_{\mathcal{M}^2}^2 = E\left[\left(\int_0^T \vartheta_t \, dM_t\right)^2\right] = E\left[\int_0^T \vartheta_t^2 \, d[M]_t\right] = \|\vartheta\|_{L^2(M)}^2,$$

hence $\mathcal{S}(M)$ is isometrically isomorphic to $L^2(M)$ and therefore complete. Closedness follows since a subspace of a Hilbert space is complete if and only if it is closed. The statement about stopping is obvious. □

Definition 2.30. The process $M \in \mathcal{M}_{loc}^2$ has the **predictable representation property** (PRP) if $\mathcal{S}(M) = \mathcal{M}^2$. That is, every $N \in \mathcal{M}^2$ can be written as $N = N_0 + \int \vartheta \, dM$ where $\vartheta \in L^2(M)$.

As a word of warning: it is essential to this definition that $\vartheta \in L^2(M)$. In fact, a ramification of the following result has been shown by Emery *et al.* (1983): Let $\mathbb{F}$ be a filtration such that all $\mathbb{F}$-adapted martingales are continuous, and H an $\mathcal{F}_T$-measurable random variable. If M is a continuous local $\mathbb{F}$-martingale with no intervals of constancy then there exists a $\vartheta \in L(M)$ such that $H = \int_0^T \vartheta_t \, dM_t$. In particular, there exists a $\psi \in L(M)$ such that

$$1 = \int_0^T \psi_t \, dM_t.$$

Of course, ψ cannot be in $L^2(M)$ since then $E\left[\int_0^T \psi_t \, dM_t\right] = 0$. In general the PRP does not hold. Instead we have the following orthogonal decomposition of L^2-martingales.

Theorem 2.31. *Kunita-Watanabe (KW) decomposition.* *Let $M \in \mathcal{M}_{loc}^2$, and $N \in \mathcal{M}^2$. Then there exists uniquely a decomposition*

$$N = N_0 + \int \vartheta \, dM + L \tag{2.4}$$

where $\vartheta \in L^2(M)$, and $L \in \mathcal{M}_0^2$ is strongly orthogonal to all elements of $\mathcal{S}(M)$. In particular, we have $\langle M, L \rangle = 0$.

Proof. By Proposition 2.29, $\mathcal{S}(M)$ is a closed subspace of the Hilbert space $\mathcal{M}^2$. Therefore, by orthogonal projection, there exists uniquely an $L \in \mathcal{S}(M)^\perp$ such that (2.4) holds. Here the superscript $\perp$ refers to the orthogonal complement where orthogonality is defined in terms of the inner product in $\mathcal{M}^2$. We still have to show that L is even strongly orthogonal to all elements in $\mathcal{S}(M)$. For this we fix an arbitrary stopping time $\tau \leq T$ and a $\vartheta \in L^2(M)$. By the tower property we obtain

$$E\left[L_\tau \int_0^\tau \vartheta_t \, dM_t\right] = E\left[L_T \int_0^\tau \vartheta_t \, dM_t\right]$$

$$= E\left[L_T \int_0^T \mathbb{I}_{[0,\tau]}(t) \, \vartheta_t \, dM_t\right]$$

$$= 0,$$

where the last equality follows from $L \in \mathcal{S}(M)^\perp$ together with $\mathbb{I}_{[0,\tau]}(.)\, \vartheta \in L^2(M)$. By Proposition 2.27$(ii)$-$(iii)$, L and $\int \vartheta \, dM$ are indeed strongly orthogonal, and $\langle \int \vartheta \, dM, L \rangle = 0$. Choosing $\vartheta = 1$ gives the last assertion, since $M \in \mathcal{M}_{loc}^2$, $L \in \mathcal{M}^2$ implies that $\langle M, L \rangle$ is well-defined. $\qquad\square$

By localisation, we get the following result:

Theorem 2.32. *KW decomposition, local version.* *Let $M, N \in \mathcal{M}^2_{loc}$.*
Then there exists uniquely a decomposition

$$N = N_0 + \int \vartheta \, dM + L$$

where ϑ is locally in $L^2(M)$, and $L \in \mathcal{M}^2_{loc}$ with $L_0 = 0$ is such that $\langle M, L \rangle = 0$.

This decomposition has been generalised by Galtchouk to the multi-dimensional case and is accordingly called the Galtchouk-Kunita-Watanabe decomposition (in short: GKW decomposition). See Jacod (1979), Chapter IV, Section 2 for a proof. We will have only need for a version for a continuous martingale integrator.

Theorem 2.33. *Galtchouk-Kunita-Watanabe decomposition.* *Let $M = (M^1, \ldots, M^d)'$ be a continuous $\mathbb{R}^d$-valued square-integrable martingale with pairwise orthogonal components and N a real-valued square-integrable martingale. Then there exists $\vartheta \in L^2(M)$ as well as $L \in \mathcal{M}^2_0$ with $[M^i, L] = 0$ for $i = 1, \ldots, d$ such that*

$$N = N_0 + \int \vartheta' \, dM + L.$$

We conclude this section with examples of stochastic processes which have the PRP with respect to their natural augmented filtrations:

(A) Brownian motion B has the PRP; consequently, all martingales with respect to the filtration generated by B are continuous.

(B) The compensated Poisson process has the PRP.

(C) More generally, **Azéma martingales** which are solutions in a weak sense to:

$$d[X, X]_t = dt + (\alpha + \beta X_{t-}) dX_t$$

where α, β are constants, have the PRP. The case $\alpha = \beta = 0$ corresponds to Brownian motion, whereas $\alpha = 1$ and $\beta = 0$ corresponds to the compensated Poisson process with intensity one. However, there are many more examples of Azéma martingales, see Protter (2005), Section IV.5 for a discussion.

Finally, we present without proof a deep result by Yor (1978) which is essential to our discussion of the minimal entropy martingale measure. Its proof crucially relies on results about BMO-martingales which are beyond the scope of this book.

Theorem 2.34. L^1**-martingale representation.** *Let M be a locally bounded local martingale, and $(\vartheta^{(n)}) \subset L(M)$ a sequence such that for each $\vartheta^{(n)}$, $\int \vartheta^{(n)} \, dM$ is a martingale on $[0, T]$. If $\int_0^T \vartheta_t^{(n)} \, dM_t$ converges in $L^1(P)$ to a random variable $\xi \in L^1(P)$ then there exists $\vartheta \in L(M)$ such that $\int \vartheta \, dM$ is a martingale and $\xi = \int_0^T \vartheta_t \, dM_t$.*

2.4 Change of measure

In finance it is often important to consider different probability measures. The statistical measure, commonly denoted by P, is supposed to reflect the real-world dynamics of financial assets. In contrast, the measure of choice for the valuation of derivatives, often denoted Q, is a so-called risk-neutral measure. Traded assets are supposed to be (local) Q-martingales, hence their dynamics under Q typically differs from their behaviour modelled under P. How much can the qualitative behaviour of a traded asset's dynamics reasonably be allowed to differ with respect to these two measures? We would like to ensure that events which have P-probability zero, like a stock price exploding to infinity, also have zero Q-probability in the risk neutral world. This discussion leads to the notion of absolute continuity.

Definition 2.35. Let P, Q be two probability measures defined on a measurable space $(\Omega, \mathcal{F})$. The measure Q is **absolutely continuous** with respect to P, denoted by $Q \ll P$, if all P-zero sets are also Q-zero sets. If $Q \ll P$ and $P \ll Q$ then P and Q are **equivalent**, denoted by $P \sim Q$. In other words, when two measures are equivalent, they have the same zero sets.

Let $Q \ll P$. By the Radon-Nikodym theorem there exists a **density** $Z = dQ/dP \in L^1(P)$ so that for $f \in L^1(Q)$ we can calculate its expectation with respect to Q by

$$E_Q[f] = E_P[Zf].$$

Note that if Q is absolutely continuous, but not equivalent to P, then we have $P(Z = 0) > 0$.

Assume that we have a filtration $\mathbb{F}$ at our disposal to examine the dynamic case. Let for $t \le T$

$$Z_t = E_P\left[Z \mid \mathcal{F}_t\right].$$

The martingale $Z = (Z_t)$ is referred to as the **density process** of Q. To calculate conditional expectations with respect to Q in terms of P, we may use **Bayes' formula**. Let $0 \le s \le t \le T$ and let $f \in L^1(Q)$ be $\mathcal{F}_t$-measurable. Then

$$Z_s E_Q\left[f \mid \mathcal{F}_s\right] = E_P\left[Z_t f \mid \mathcal{F}_s\right].$$

A consequence of Bayes' formula is that M is a Q-martingale if and only if ZM is a P-martingale. Hence, any Q-martingale can be turned into a P-martingale by multiplying it with the density process. Note that the martingale property itself is *not* invariant under equivalent measure changes.

However, there are important objects like stochastic integrals and quadratic variations which are invariant under equivalent measure changes, even though they depend by definition on a probability measure P. This dependence is only via the P-null sets, so they could be defined equally well under any equivalent probability measure.

In contrast, the angle bracket process is *not* invariant under equivalent measure changes because it depends by definition on the martingale property. Recall that it is defined (for locally square-integrable semi-martingales S) as the process $\langle S \rangle$ one has to subtract from the quadratic variation process $[S]$ to turn it into a local martingale. Since the martingale property typically gets lost when switching the measure, we would expect to have to adjust the choice of $\langle S \rangle$. For example, consider a Poisson process N with intensity λ. We have $[N] = N$, so the angle bracket equals λt. As we shall see below, the effect of an equivalent measure change is that the intensity changes, so the angle bracket under the new measure would be μt for some $\mu \ne \lambda$.

Although the martingale property is not preserved under measure changes, fortunately at least the semi-martingale property is preserved. Moreover, it is possible to state the precise semi-martingale decomposition under the new measure Q. This result is known in the literature as 'Girsanov's theorem'.

Let us first give two examples. They both are consequences of the general formulation of Girsanov's theorem to be given below.

(A) Let B be a P-Brownian motion, $\mu \in \mathbb{R}$, and define an equivalent measure Q via its density

$$\frac{dQ}{dP} = \exp\left(-\mu B_T - \frac{1}{2}\mu^2 T\right).$$

Then $\widehat{B} = B + \mu t$ is a Q-Brownian motion (up to time T). Alternatively stated, the semi-martingale decomposition of B under Q is $B = \widehat{B} - \mu t$. Hence the effect of the measure change is to add a drift term to the Brownian motion.

(B) Let $N - \lambda t$ be a compensated Poisson process with P-intensity $\lambda > 0$, and let $\kappa > -1$. Define an equivalent measure Q by

$$\frac{dQ}{dP} = e^{-\kappa \lambda t} \prod_{0 < s \le t} (1 + \kappa \Delta N_s)$$

$$= e^{-\kappa \lambda t} (1 + \kappa)^{N_t}$$

$$= \exp\left(N_t \log\left(1 + \kappa\right) - \kappa \lambda t\right).$$

Then N is a Poisson process under Q with intensity $(1 + \kappa)\lambda$. The process $N - (1 + \kappa)\lambda t$ is a compensated Poisson process under Q and thus a Q-martingale. The effect of the measure change is to change the intensity of the Poisson process, or in other words, to add a drift term to the compensated Poisson process.

Let us now state a general form of Girsanov's theorem. It is not the most general setting, as we will assume that Q is equivalent to P which suffices for most applications in finance. There is a version for the case $Q \ll P$ due to Lenglart which will be discussed in Section 3.6.

Theorem 2.36. *Girsanov's theorem, standard version.* *Let $Q \sim P$, with density process given by*

$$Z_t = E\left[\frac{dQ}{dP}\middle| \mathcal{F}_t\right].$$

If S is a semi-martingale under P with decomposition $S = M + A$ where M is a local P-martingale, and A a FV-process, then S is a semi-martingale under Q and has a decomposition

$$S = \left(M - \int \frac{1}{Z} d\left[Z, M\right]\right) + \left(A + \int \frac{1}{Z} d\left[Z, M\right]\right) \qquad (2.5)$$

where $M - \int \frac{1}{Z} d\left[Z, M\right]$ is a local Q-martingale.

In situations where the process S may exhibit jumps, it is often more convenient to apply a version of Girsanov which uses the angle bracket instead of the quadratic co-variation. In this predictable version a local P-martingale is transformed into a *special* Q-semi-martingale with a unique canonical decomposition.

Theorem 2.37. *Girsanov's theorem, predictable version.* *Let $Q \sim P$, with density process as above, and let $S = M + A$ be a P-semi-martingale. Given that $\langle Z, M \rangle$ exists (with respect to P), then the canonical decomposition of S under Q is*

$$S = \left(M - \int \frac{1}{Z_-} \, d \langle Z, M \rangle \right) + \left(A + \int \frac{1}{Z_-} \, d \langle Z, M \rangle \right). \qquad (2.6)$$

One of the most important applications of measure changes in mathematical finance is to find martingale measures for the price process S of a risky asset.

Definition 2.38. An **equivalent martingale measure** for S is a probability measure $Q \sim P$ such that S is a Q-local martingale.

Consider for example the Bachelier model where $S = B + \mu t$ is a Brownian motion plus drift. If we take as above the measure change as given by a density process $Z = \exp \left(-\mu B - \frac{1}{2} \mu^2 t \right)$ then we have (since $dZ = -\mu Z \, dB$)

$$A + \int \frac{1}{Z} \, d \, [Z, M] = \mu t + \int \frac{1}{Z} \, d \left[-\mu \int Z \, dB, B \right]$$

$$= \mu t + \int \frac{1}{Z} \, d \left(-\mu \int Z \, dt \right)$$

$$= 0.$$

In this case the standard version of Girsanov's theorem coincides with the predictable one since S is continuous. According to Girsanov's theorem the price process S is a Q-local martingale and then a Brownian motion due to Lévy's characterisation. Therefore Q is a martingale measure for S.

Let us now state an important structural result concerning the price process S with semi-martingale decomposition $S = M + A$ when an equivalent martingale measure Q exists. Firstly, let Z denote the density process of Q with respect to P and assume that $\langle Z, M \rangle$ exists, then the predictable version of Girsanov's theorem implies that to ensure that S is a local Q-martingale we must have that

$$A = - \int \frac{1}{Z_-} \, d \langle Z, M \rangle.$$

Next associate with the FV-process $\langle Z, M \rangle$ (resp. $\langle M \rangle$) a measure on the non-negative real line, denoted $d\langle Z, M \rangle$ (resp. $d\langle M \rangle$). It follows from the Kunita-Watanabe inequality that

$$d\langle Z, M \rangle \ll d\langle M \rangle,$$

and hence that

$$dA \ll d\langle M \rangle.$$

Consequently there exists some predictable process λ such that

$$S = M + \int \lambda \, d\langle M \rangle. \tag{2.7}$$

In that case, the remarkable conclusion we can draw from (2.7) is that the existence of an equivalent martingale measure for S implies that S is a special semi-martingale, i.e. its finite variation part is predictable and therefore the semi-martingale decomposition (2.7) is unique.

Definition 2.39. We say that the **structure condition** holds if there exists $M \in \mathcal{M}^2_{loc}$ and a predictable process λ such that

$$S = M + \int \lambda \, d\langle M \rangle \tag{2.8}$$

and

$$\int_0^T \lambda_t^2 \, d\langle M \rangle_t < \infty. \tag{2.9}$$

This section concludes with two examples of the decomposition (2.8).

(A) Geometric Brownian motion

Let B be a standard Brownian motion and recall that $\langle B \rangle = [B] = t$ since Brownian motion is a continuous process. Consider the following dynamics for the price process S:

$$dS = \mu S \, dt + \sigma S \, dB.$$

Here μ, $\sigma > 0$ are constants. The martingale part of these dynamics is $dM = \sigma S \, dB$, which has angle bracket process $d\langle M \rangle = \sigma^2 S^2 \, dt$. Therefore, combining equation (2.8) with the price dynamics yields

$$\lambda = \frac{\mu}{\sigma^2 S}.$$

(B) Compensated Poisson process

Let N be a Poisson process with intensity $\gamma > 0$ and consider the following dynamics for the price process S:

$$dS = \mu \, dt + \rho \, dN.$$

Here $\mu < 0$, $\rho > 0$ are constants. Since $N - \gamma t$ is a martingale, the martingale part of the dynamics is $M = \rho (N - \gamma t)$. In which case, $\langle M \rangle = \rho^2 \langle N \rangle = \rho^2 \gamma t$ and it follows from equation (2.8) that

$$\lambda = \frac{\mu + \rho \gamma}{\rho^2 \gamma}.$$

2.5   Stochastic exponentials

Stochastic exponentials have two main roles in financial mathematics: firstly they are candidates for the density process described in the previous section and secondly they can be used to model asset prices under the risk-neutral measure Q. Crucially however, given that X is a martingale, the stochastic exponential $\mathcal{E}(X)$ may fail to be a martingale. This section examines some criteria required to ensure that $\mathcal{E}(X)$ is indeed a martingale. The following 'definition' in fact contains a statement which easily follows from a standard existence and uniqueness result for linear stochastic integral equations.

Definition 2.40. Let X be a semi-martingale with $X_0 = 0$. Then, there exists a unique semi-martingale Z that satisfies the equation

$$Z = 1 + \int Z_- \, dX.$$

The process Z is called the **stochastic exponential** of X and is denoted by $\mathcal{E}(X)$.

We begin this section with some examples:

(A) If B is a Brownian motion, then an application of Itô's formula reveals that

$$\mathcal{E}(B) = \exp\left(B - \frac{1}{2}t\right).$$

(B) The stochastic exponential of a compensated Poisson process $N - \gamma t$ is

$$\mathcal{E}\left(N - \gamma t\right) = \exp\left(-\frac{1}{2}\gamma t\right) \times 2^N$$

$$= \exp\left(N \log(2) - \frac{1}{2}\gamma t\right).$$

(C) The classical Samuelson model for the evolution of stock prices is given as a stochastic exponential. The price process S is modelled here as solution of the stochastic differential equation

$$dS = \mu\, S\, dt + \sigma S\, dB.$$

In the case of constant drift coefficient μ and volatility $\sigma > 0$ the solution to this equation is

$$S = \mathcal{E}\left(\sigma B + \mu t\right)$$

$$= \exp\left(\sigma B + \left(\mu - \frac{1}{2}\sigma^2\right)t\right).$$

The general situation is covered by the following result.

Proposition 2.41. *Stochastic exponential.* *For a semi-martingale X, the stochastic exponential is given as*

$$\mathcal{E}\left(X\right)_t = \exp\left(X_t - \frac{1}{2}[X^c]_t\right) \prod_{0 < s \leq t} \left(1 + \Delta X_s\right) e^{-\Delta X_s}$$

where $[X^c]_t = [X]_t - \sum_{s \leq t} (\Delta X_s)^2$, and the possibly infinite product converges. In the case when X is a local martingale vanishing at zero with $\Delta X > -1$, the process $\mathcal{E}\left(X\right)$ is a strictly positive local martingale.

Given that the structure condition holds, the following result shows that density processes of martingale measures are given as stochastic exponentials, parametrised by some local martingale L.

Theorem 2.42. *Structure of martingale measures.* *Let S be a semi-martingale which fulfills the structure condition.*

(i) Let Q be an equivalent martingale measure for S. Then the density process Z of Q with respect to P is given by the stochastic exponential

$$Z = \mathcal{E}\left(-\int \lambda\, dM + L\right) \tag{2.10}$$

for some local P-martingale L such that $[M, L]$ is a local P-martingale.

(*ii*) *Let $Q \sim P$ be a probability measure whose density process can be written like in (2.10) where L and $[M, L]$ are local P-martingales. Then Q is an equivalent martingale measure for S.*

Proof. (*i*) Since $Q \sim P$, the density process Z is strictly positive and we may write $Z = \mathcal{E}(N)$ for some local martingale N with $\Delta N > -1$. The structure condition implies that $\int \lambda \, dM \in \mathcal{M}_{loc}^2$. Define the local P-martingale L by

$$L := N + \int \lambda \, dM.$$

We have to show that $[M, L]$ is a local P-martingale. By integration by parts,

$$SZ = \int S_- \, dZ + \int Z_- \, dS + [S, Z].$$

Since S is a local Q-martingale, SZ is a local P-martingale, and by Theorem 2.24 so is $\int S_- \, dZ$. It follows that $\int Z_- \, dS + [S, Z]$ is a local P-martingale, and, by integrating $1/Z_-$ with respect to it and applying Theorem 2.24, finally we get that $S + \int 1/Z_- \, d[S, Z]$ is a local P-martingale. We have, by the stochastic integral equation satisfied by Z,

$$S + \int \frac{1}{Z_-} \, d[S, Z] = M + \int \lambda \, d\langle M \rangle + \int \frac{1}{Z_-} \, d\left[S, \int Z_- \, dN \right]$$

$$= M + \int \lambda \, d\langle M \rangle + [S, N]$$

$$= M + \int \lambda \, d\langle M \rangle + [M, N] + \left[\int \lambda \, d\langle M \rangle, N \right]$$

$$= M - \int \lambda \, d([M] - \langle M \rangle) + [M, L] + \left[\int \lambda \, d\langle M \rangle, N \right].$$

Here the last equality follows, since by the Kunita-Watanabe inequality and the structure condition, $\int |\lambda| \, d\langle M \rangle$ and therefore also $\int |\lambda| \, d[M]$ have locally integrable variation. This implies that

$$\int \lambda \, d[M] - \int \lambda \, d\langle M \rangle = \int \lambda \, d([M] - \langle M \rangle)$$

is a local P-martingale. Moreover, $[\int \lambda \, d\langle M \rangle, N]$ is a local P-martingale by Theorems 2.20 and 2.24. Therefore, $[M, L]$ can be written as sum of local P-martingales which proves the claim.

(*ii*) We have to show that SZ is a local P-martingale. By integration by parts,

$$SZ = \int S_- \, dZ + \int Z_- \, dS + [S, Z]$$

$$= \int S_- \, dZ + \int Z_- \, dM + \int Z_- \lambda \, d \langle M \rangle$$

$$+ \left[M + \int \lambda \, d \langle M \rangle, \int Z_- \, d \left(- \int \lambda \, dM + L \right) \right]$$

$$= \int S_- \, dZ + \int Z_- \, dM + \int Z_- \, d [M, L]$$

$$- \int Z_- \lambda \, d \left([M] - \langle M \rangle \right) + \left[\int \lambda \, d \langle M \rangle, \int Z_- \, d \left(- \int \lambda \, dM + L \right) \right].$$

The first three terms are local P-martingales by Theorem 2.24. The fourth term is a local P-martingale by the same argument as under (*i*). Finally, the fifth term is a local P-martingale as well by Theorems 2.20 and 2.24. □

Stochastic exponentials are intrinsically related to measure changes since they qualify as candidates for density processes. When the stochastic exponential is positive, it is possible to define a new measure Q on $\mathcal{F}_T$ via

$$\frac{dQ}{dP} = Z_T.$$

If Z is a martingale, then Q is a probability measure since $E[Z_T] = Z_0 = 1$. However, if Z is a positive strict local martingale then it is a strict super-martingale and consequently we have $Q(\Omega) = E[Z_T] < 1$ so Q is no longer a probability measure. For this reason it is of paramount interest to have criteria available to validate whether stochastic exponentials are martingales.

Theorem 2.43. *Protter & Shimbo's criterion.* Let M be a locally square integrable martingale such that $\Delta M > -1$. If

$$E \left[e^{\frac{1}{2} \langle M^c \rangle_T + \langle M^d \rangle_T} \right] < \infty \qquad (2.11)$$

where M^c and M^d are the continuous and purely discontinuous martingale parts of M, then $\mathcal{E}(M)$ is a strictly positive martingale on $[0, T]$.

If M is continuous, $\langle M^c \rangle = [M]$ and (2.11) is known as **Novikov's condition**. Another general criterion can be formulated using BMO-theory.

Definition 2.44. Let M be a martingale with $M_0 = 0$. M is a **BMO-martingale** if there exists a constant C such that for all stopping times $\tau \in [0, T]$

$$E\left[\|M_T - M_{\tau-}\| \, | \, \mathcal{F}_\tau\right] \leq C.$$

BMO-martingales respectively their quadratic variations enjoy the existence of certain exponential moments.

Proposition 2.45. *John-Nirenberg inequality.* Let M be a *BMO*-martingale. Then there exists $\varepsilon > 0$ such that

$$E\left[e^{\varepsilon[M]_T}\right] < \infty.$$

For the following result, see Theorem 2.3 and Remark 2.3 in Kazamaki (1994).

Theorem 2.46. *BMO-criterion.* Let M be a *BMO*-martingale satisfying $\Delta M \geq -1 + \delta$ for some δ with $0 < \delta \leq 1$. Then $\mathcal{E}(M)$ is a martingale.

2.6 Notes and further reading

Most of this chapter is based on Protter (2005) which serves as the basic reference for stochastic calculus for this book. Other books on a general level include Dellacherie & Meyer (1980), He *et al.* (1992), Jacod & Shiryaev (2003), and Rogers & Williams (2000). Parts of the sections about measure changes and stochastic exponentials have been published in similar form by the first author in the Encyclopedia of Quantitative Finance (2010). The structure condition is due to Ansel & Stricker (1992), and Theorem 2.42 is adapted from Steiger (2005). Protter & Shimbo's criterion is due to, of course, Protter & Shimbo (2008). While we touch on *BMO*-theory only lightly, it is essential for proving many of the deeper results instrumental for hedging in incomplete markets, for example L^1-martingale representation as in Theorem 2.34. See Kazamaki (1994) for further reading.

Chapter 3

Arbitrage and Completeness

This chapter introduces the basic concepts of arbitrage and completeness, and discusses some aspects of the two fundamental theorems of asset pricing. Moreover, we study situations where the structure condition for the asset price process fails to hold, and give examples of the resulting arbitrage opportunities. Finally, the optional decomposition theorem is related to the concept of super-hedging, and is shown to be instrumental in discussing arbitrage in models where certain constraints are imposed on the price process.

3.1 Strategies and arbitrage

We model the price process of a risky asset by a semi-martingale denoted S. The savings account, R, is assumed to be a continuous, strictly positive process of finite variation. A prime example would be

$$R_t = \exp\left(\int_0^t r_s \, ds\right)$$

where the short rate r is an adapted integrable process. In general, a strategy is a pair (ϑ, ψ) where the predictable, S-integrable process ϑ is the number of units of the risky asset held in the portfolio and ψ is an adapted R-integrable process describing the holdings in the savings account. The value of this portfolio at time t equals

$$V_t = \vartheta_t \, S_t + \psi_t \, R_t,$$

and a strategy is called **self-financing** if for all $t \geq 0$

$$V_t = V_0 + \int_0^t \vartheta_u \, dS_u + \int_0^t \psi_u \, dR_u.$$

By choosing R as a **numeraire**, it is possible to reduce this situation to one where the savings account is trivial ($R = 1$). This is achieved by dividing all financial quantities by R. An application of the integration by parts formula shows that the strategy (ϑ, ψ) for (S, R) is self-financing if and only if (ϑ, ψ) is self-financing for $(S/R, 1)$. Hence for a self-financing strategy, we only need to specify ϑ and the initial capital V_0. As $dR = 0$ for $R = 1$, the self-financing constraint then determines the value process V and the bank account holdings uniquely. In the remainder of this chapter we will always use this technique to work with the discounted price process together with the trivial savings account (for which $dR = 0$). In particular, with a slight abuse of notation, from here on we denote S/R by S without further mention.

Definition 3.1. A **strategy** ϑ is an S-integrable process. The **value process** $V = V(c, \vartheta)$ associated to an initial capital c and a strategy ϑ in the risky asset S is given as the stochastic integral process

$$V = c + \int \vartheta \, dS.$$

Here ϑ_t can be interpreted as the number of shares held at time t, while the stochastic integral process $\int \vartheta \, dS$ is the outcome from trading in S, often optimistically called the 'gains process'.

Definition 3.2. A probability measure Q absolutely continuous with respect to P is a **martingale measure** for S if S is a Q-local martingale. It is called an **equivalent martingale measure** if it is also equivalent to P. Let $\mathcal{M}$ denote the set of all martingale measures for S and $\mathcal{M}^e$ the set of all equivalent martingale measures for S.

In many cases, the associated value process V is a Q-local martingale for every martingale measure Q such that V is Q-integrable. This holds for instance when the strategy ϑ is locally bounded or when V is a special semi-martingale. However, it is not true in full generality and has led to the concept of σ-martingales; see Delbaen & Schachermayer (2006). Ensuring that V is a Q-local martingale does not necessarily exclude arbitrage phenomena as is shown by the following example. For simplicity, we will work with an infinite time horizon although this is not essential.

Example 3.3. Suicide strategy.

Assume that the price process of a risky asset evolves as the stochastic exponential $Z = \exp\left(B - \frac{1}{2}t\right)$ where B is a standard Brownian motion starting in zero. As an exception to our usual rule, Z is defined on the whole of $\mathbb{R}_+$. Since one-dimensional Brownian motion is almost-surely recurrent, zero must be an accumulation point of Z. As Z can be written as a stochastic integral of B, it is a positive local martingale, and hence a super-martingale by Proposition 2.6. We conclude via the super-martingale convergence theorem that

$$\lim_{t \to \infty} Z_t = 0 \qquad P - \text{a.s.}$$

Investing in the asset with price process Z therefore amounts to following a **suicide strategy**, since one starts with an initial capital of one and ends up with no money at time infinity. The mathematical explanation for this phenomenon is that Z is not a martingale on the closed interval $[0, \infty]$, or equivalently, the family $\{Z_t, \ t \in \mathbb{R}_+\}$ is not uniformly integrable. This example can be modified to generate an **arbitrage opportunity**: By the equation satisfied by the stochastic exponential,

$$V_t := \int_0^t (-Z_s) \, dB_s = 1 - Z_t \to 1 \quad \text{for } t \to \infty.$$

Hence investing into a stock whose price process follows a Brownian motion via the self-financing strategy $-Z$ generates a sure profit of one unit (albeit after an infinite amount of time), but requires no initial capital. However, it is not recommendable to pursue such an arbitrage strategy, at least not if the agent has a finite credit line, since one can show that

$$E \left[\inf_{t > 0} V_t \right] = -\infty.$$

As shown by this example, it is necessary to restrict ourselves to considering admissible strategies which exclude the pathological behaviour just described.

Definition 3.4. Assume that $\mathcal{M}^e \neq \varnothing$. A strategy ϑ is an **admissible strategy** if the associated value process V is a Q-super-martingale for every martingale measure $Q \in \mathcal{M}^e$.

If the value process V is almost surely bounded from below, then it follows from Fatou's lemma that the associated strategy ϑ is admissible. We can interpret this condition as a situation where the agent has a finite credit line. However, we shall need sometimes a more restrictive definition of an admissible strategy (for example in the next section about complete markets, or in the context of mean-variance hedging), and will make this then always explicit. *The definition of admissibility is specific to each section and will be revisited when necessary.*

Intuitively an arbitrage opportunity is a way to generate a riskless profit by investing in some risky asset. It is commonly assumed that financial markets are 'efficient' in the sense that there are no arbitrage opportunities. Note that the probability measure P only enters the following definition of an arbitrage opportunity through the null sets of P.

Definition 3.5. A strategy ϑ is called an **arbitrage opportunity** if we have for the associated value process V that

(i) $V_0 \leq 0$

(ii) $V_T \geq 0 \quad P-\text{a.s.}$

(iii) $P(V_T > 0) > 0$.

The **first fundamental theorem of asset pricing** states that there is 'No Free Lunch with Vanishing Risk' *if and only if* there exists an equivalent martingale measure for S. 'No Free Lunch with Vanishing Risk' is a technical condition related to the absence of arbitrage which allows one to start from an exact economic condition and then to look for possible mathematical implications; see Delbaen & Schachermayer (2006). However, in this book we typically start with some concrete mathematical model and have just to make sure that it does not allow for arbitrage.

Theorem 3.6. *Easy direction of first fundamental theorem of asset pricing. If there exists an equivalent martingale measure Q for S then there are no arbitrage opportunities with admissible strategies.*

Proof. Let ϑ be an admissible arbitrage opportunity and V its associated value process. Since $Q \ll P$ we then also have

$$V_T \geq 0 \qquad Q - \text{a.s.}$$

As V is a Q-super-martingale by our definition of admissibility, we have

$$E_Q[V_T] \leq V_0 \leq 0,$$

hence

$$Q(V_T > 0) = 0.$$

As $P \ll Q$ it follows that

$$P(V_T > 0) = 0,$$

contradicting that ϑ is an arbitrage opportunity. $\square$

3.2   Complete markets

In a complete market, all claims are 'redundant' in the sense that they can be replicated by the outcome from trading with a self-financing admissible strategy in the underlying asset. The initial capital needed to drive such a replicating strategy then corresponds to the fair price of the claim. The standing assumption for this section is

$$\mathcal{M}^e \neq \varnothing.$$

Definition 3.7. A **claim** is an $\mathcal{F}_T$-measurable random variable. The claim H is **attainable** if there exist a constant c and an admissible strategy ϑ such that

$$H = c + \int_0^T \vartheta_t \, dS_t.$$

Here a strategy ϑ is **admissible** if and only if the stochastic integral $\int \vartheta \, dS$ is a Q-martingale for all $Q \in \mathcal{M}^e$. The quintuple $(\Omega, \mathcal{F}, \mathbb{F}, P, S)$ is called a **market**. A market is **complete** if all bounded claims are attainable. Finally, a market which is not complete is called **incomplete**.

We will now discuss the **second fundamental theorem of asset pricing** which states that an arbitrage-free market is complete *if and only if* the equivalent martingale measure is unique.

Let us fix some equivalent martingale measure Q for S and consider for each bounded claim H the associated Q-martingale M given by

$$M_t = E_Q \left[H \mid \mathcal{F}_t \right], \qquad t \leq T.$$

To identify a complete market we need to be able to tell whether it is possible to represent M as a stochastic integral with respect to S. We will certainly be able to decompose M uniquely into a stochastic integral with respect to S and a strongly orthogonal local martingale L, corresponding to an attainable and non-attainable part, see Theorem 2.32. Nevertheless, to identify attainable claims we need the predictable representation property (PRP) to hold. We are now ready to state the proposed characterisation of complete markets, and give a proof of the easy implications.

Theorem 3.8. *Second fundamental theorem of asset pricing. The following assertions are equivalent:*

(*i*) *The market is complete.*

(ii) $|\mathcal{M}^e| = 1$, *i.e.* $\mathcal{M}^e$ *consists of a singleton.*
(iii) *There exists* $Q \in \mathcal{M}^e$ *such that* S *has the PRP with respect to* $(Q, \mathbb{F})$.

Proof. $(i) \Rightarrow (ii)$ Take $\Gamma \in \mathcal{F}_T$. Since the market is complete, there exists a constant c and an admissible strategy ϑ such that

$$\mathbb{I}_\Gamma = c + \int_0^T \vartheta_t \, dS_t.$$

For any $Q \in \mathcal{M}^e$ we have, by the admissibility of ϑ,

$$Q(\Gamma) = E_Q[\mathbb{I}_\Gamma] = c.$$

Therefore, Q is unique.

$(ii) \Rightarrow (iii)$ This is the difficult part, for which we refer to Protter (2005), Section IV.3.

$(iii) \Rightarrow (i)$ Take any bounded $\mathcal{F}_T$-measurable claim H and consider the associated Q-martingale

$$M_t = E_Q[H \mid \mathcal{F}_t].$$

By the PRP,

$$M = M_0 + \int \vartheta \, dS.$$

The boundedness of M implies that ϑ is an admissible strategy by Theorem 2.15.   □

This theorem shows that complete markets are characterised by price processes that have the PRP. As a consequence markets containing a single price process S, such that S is a diffusion driven by a one dimensional Brownian motion B and have $\mathbb{F}^S = \mathbb{F}^B$, are complete. It is important to note that these are not the only markets that are complete. In particular, purely discontinuous price processes, such as Azéma martingales can lead to complete markets. For more details on processes that exhibit the PRP see Section 2.3.

Let us now consider a bounded claim H in a complete market. The assumption of boundedness is imposed in order to avoid certain technicalities, however, this assumption is generally not too restrictive. For example, despite the payoff of a call option being unbounded, we can express its payoff via the put-call parity in terms of a future and the bounded payoff of a put option.

As the market is complete, H is attainable, and hence there exist a constant c and an admissible strategy ϑ such that

$$H = c + \int_0^T \vartheta_t \, dS_t.$$

The constant c is the initial capital required to drive the self-financing replicating strategy ϑ. As the sum of initial capital and outcome from trading matches exactly the payoff of the claim H, it is reasonable to call c the **fair price** of the claim H.

On the other hand, by the second fundamental theorem of asset pricing, there exists a unique martingale measure Q. By taking the Q-expectation of both sides of the equation, and noting the admissibility of ϑ (hence $\int \vartheta \, dS$ is a Q-martingale) we obtain that

$$c = E_Q[H].$$

In conclusion, the fair price of the claim H can be calculated by taking the expectation of the payoff H under the martingale measure Q. The replicating strategy ϑ is called the **hedging strategy** for the claim H.

3.3 Hidden arbitrage and local times

The purpose of this section is to give an example of a market where all claims are attainable in a sense to be specified, yet arbitrage opportunities exist. Moreover, it provides a concrete arbitrage strategy when the price process S cannot be written in its canonical form (2.8). We assume that the process S is continuous and is given by

$$S = M + \int \lambda \, d[M] + L \tag{3.1}$$

where λ is predictable, M is a continuous local martingale and L is an adapted, continuous and positive non-zero process of finite variation such that the induced measure L is singular with respect to $[M]$. The continuity of S implies that $[M] = \langle M \rangle$ so L is also singular with respect to $\langle M \rangle$. An example of a suitable process L would be the local time at the level c of the martingale M, denoted L^c. Such a local time process L^c is increasing only when the martingale M hits the level c, and intuitively speaking, it measures the amount of time the martingale M spends at the level c. For a discussion of local time and related results, see Protter (2005), Section IV.7.

Theorem 3.9. *Local time.* *Let X be a semi-martingale, c a constant, and (π_n) be a sequence of partitions of $[0,t]$ whose grid size $\|\pi_n\|$ tends to zero. The limit*

$$L_t^c = 2 \lim_{\|\pi_n\| \to 0} \sum_i (c - X_{t_i+1})\, \mathbb{I}_{\left\{ sign\left(X_{t_{i+1}-c}\right) < sign\left(X_{t_i-c}\right) \right\}}$$

*exists uniformly on compact time sets in probability and is called the **local time** of X at the level c. L^c is a finite variation process where the induced measure is singular with respect to $d\,[X]$.*

In case of price processes of the form given in equation (3.1), there does not exist an equivalent martingale measure as has been discussed in Section 2.3. Hence, by the first fundamental theorem, arbitrage opportunities may exist. A specific example is provided by the Tanaka formula,

$$S = 1 + |B| = 1 + \int sign(B)\, dB + L^0(B). \tag{3.2}$$

Here B is a Brownian motion starting in zero. The local time $L^0(B)$ is singular with respect to Lebesgue measure dt and

$$[M]_t = \left[\int_0^{\cdot} sign(B_u)\, dB_u \right]_t = [B]_t = t.$$

Obviously, there is no equivalent martingale measure since for all probability measures $Q \sim P$ we have

$$E_Q\,[S_0] = 1 < E_Q\,[S_t] \quad \text{for } t > 0.$$

This example motivates the following general result regarding a price process of the type as in (3.1).

Theorem 3.10. *Hidden arbitrage.* *Suppose there exists a unique equivalent martingale measure Q for*

$$N = M + \int \lambda\, d\,[M].$$

Then the market with continuous price process S as in (3.1) is not arbitrage-free. However, all bounded claims can be replicated by $\mathbb{F}^N$-admissible strategies. An $\mathbb{F}^N$-admissible strategy is an $\mathbb{F}^N$-predictable, N-integrable process ϑ such that $\int \vartheta\, dN$ is a Q-martingale.

Proof.   Let H be a bounded claim. By the second fundamental theorem, Theorem 3.8, there exists an $\mathbb{F}^N$-admissible strategy ϑ such that

$$H = E_Q[H] + \int_0^T \vartheta_t \, dN_t.$$

We modify ϑ by setting

$$\widetilde{\vartheta} = \vartheta \, \mathbb{I}_{\mathrm{supp}(d[M])}.$$

Since $[N] = [M]$ we still have

$$H = E_Q[H] + \int_0^T \widetilde{\vartheta}_t \, dN_t.$$

As L is singular with respect to $[M]$, we have $\int \widetilde{\vartheta} \, dL = 0$ and hence

$$H = E_Q[H] + \int_0^T \widetilde{\vartheta}_t \, dS_t.$$

Therefore, the market is complete with respect to $\mathbb{F}^N$. However, we claim that the following $\mathbb{F}^N$-predictable strategy is an arbitrage opportunity:

$$\psi = \mathbb{I}_{\mathrm{supp}(d[M])^c}.$$

To see this, first observe that as L is singular with respect to $[M]$, we have $L_T = \int_0^T \psi_t \, dL_t$. Secondly, it follows by the Itô-isometry that

$$\int_0^T \psi_t \, dN_t = 0,$$

and as a consequence

$$L_T = \int_0^T \psi_t \, dS_t.$$

Therefore, the replication of the positive, non-zero claim L_T requires no initial investment, so the market is not arbitrage free. $\square$

This phenomenon has been coined 'hidden arbitrage' since the arbitrage takes place on a time set of $d[M]$-measure zero. If M is a Brownian motion the arbitrage happens on a set of Lebesgue measure zero. In general, however, knowing about the existence of an $\mathbb{F}^N$-admissible arbitrage strategy is pretty useless unless $\mathbb{F}^N \subset \mathbb{F}^S$. Remarkably, in a Brownian setting this is indeed the case as shown by the following interesting and not obvious result.

Theorem 3.11. *Local time does not perturb the Brownian filtration.* Let

$$dN_t = \mu\left(N_t\right) dt + dB_t$$

where B is a standard Brownian motion with $B_0 = 0$ and the measurable function μ is such that there exists an equivalent probability measure Q such that N is a Q-Brownian motion. We set

$$dS_t = \mu\left(N_t\right) dt + dB_t + \alpha\, dL_t^c$$

where L^c denotes Brownian local time at the level c and $\alpha \in \mathbb{R}$. Then $\mathbb{F}^N = \mathbb{F}^S$.

Proof.   Changing to Q, it suffices to establish that $X = B + \alpha L^c$ and B generate the same filtrations. As the case for general c can be easily deduced from the case $c = 0$ we deal only with that case. We have $\mathbb{F}^X \subset \mathbb{F}^B$ since L^0 is $\mathbb{F}^B$-adapted.

It will follow that $\mathbb{F}^B \subset \mathbb{F}^X$ once we have established the existence of a random set $D \subset \mathbb{R}_+ \times \Omega$ with the following properties:

(i) The process $\mathbb{I}_D\left(\cdot, \omega\right)$ is $\mathbb{F}^X$-predictable.
(ii) $P\left(\omega : (t, \omega) \in D\right) = 1$ for all $t > 0$.
(iii) $P\left(\omega : \left(T_t\left(\omega\right), \omega\right) \in D\right) = 0$ for all $t > 0$. Here T_t denotes the right continuous inverse of L^0,

$$T_t = \inf\left\{s > 0 \mid L_s^0 > t\right\}.$$

Indeed, it follows from (ii) that

$$E\left[\int_0^\infty \mathbb{I}_{D^c}\left(s\right) ds\right] = \int_0^\infty E\left[\mathbb{I}_{D^c}\left(s\right)\right] ds = 0,$$

and therefore $\int_0^t \mathbb{I}_{D^c}\left(s\right) dB_s = 0$ and hence $\int_0^t \mathbb{I}_D\left(s\right) dB_s = B_t$. Moreover, we get from (iii) that

$$E\left[\int_0^\infty \mathbb{I}_D\left(s\right) dL_s^0\right] = E\left[\int_0^\infty \mathbb{I}_D\left(T_s\right) ds\right] = \int_0^\infty E\left[\mathbb{I}_D\left(T_s\right)\right] ds = 0,$$

hence $\int_0^t \mathbb{I}_D\left(s\right) dL_s^0 = 0$. We can therefore write

$$B_t = \int_0^t \mathbb{I}_D\left(s\right) dB_s + \alpha \int_0^t \mathbb{I}_D\left(s\right) dL_s^0 = \int_0^t \mathbb{I}_D\left(s\right) dX_s.$$

Since, by (i), $\mathbb{I}_D\left(\cdot, \omega\right)$ is $\mathbb{F}^X$-predictable, it follows that $\mathbb{F}^B \subset \mathbb{F}^X$.

Let us now construct a random set D which satisfies properties (i)–(iii) as above. We take as D all (t, ω) such that the limit

$$\lim_n \frac{1}{n} \sum_{k=1}^{n} \mathbb{I}_{\{X_{t-2^{-k}}(\omega) - X_{t-2^{-k-1}}(\omega) > 0\}}$$

exists and is almost surely equal to $1/2$.

ad (*i*): $\mathbb{I}_D(\cdot, \omega)$ is $\mathbb{F}^X$-predictable since X is continuous and, of course, $\mathbb{F}^X$-adapted.

ad (*ii*): Fix $t > 0$. With probability one, there exists $S(\omega) < t$ such that $B(\cdot, \omega)$ has no zeroes (and L^0 is hence constant) on $[S(\omega), t]$. Therefore, for k large enough we have

$$X_{t-2^{-k}} - X_{t-2^{-k-1}} = B_{t-2^{-k}} - B_{t-2^{-k-1}}.$$

It thus suffices to note that by the strong law of large numbers,

$$\lim_n \frac{1}{n} \sum_{k=1}^{n} \mathbb{I}_{\{B_{t-2^{-k}}(\omega) - B_{t-2^{-k-1}}(\omega) > 0\}} = \frac{1}{2} \qquad P-\text{a.s.}$$

ad (*iii*): Fix $t > 0$. We have to verify that

$$\lim_n \frac{1}{n} \sum_{k=1}^{n} \mathbb{I}_{\{X_{T_t - 2^{-k}}(\omega) - X_{T_t - 2^{-k-1}}(\omega) > 0\}}$$

either does not exist or is different from $\frac{1}{2}$. We now use (without proof) the fact that $X_{T_t \wedge s}$ and $\alpha t - X_{T_t - s}$ have the same law (this can be shown e.g. via excursion theory) and that it is therefore equivalent to prove that the probability of the set

$$\Gamma := \left\{ \omega \,\middle|\, \lim_n \frac{1}{n} \sum_{k=1}^{n} \mathbb{I}_{\{X_{2^{-k}}(\omega) - X_{2^{-k-1}}(\omega) > 0\}} = \frac{1}{2} \right\}$$

is zero. We proceed by contradiction: assume that $P(\Gamma) > 0$. Since $\Gamma \in \mathcal{F}_0^B$ (this uses the fact that by our convention, $\mathbb{F}^B$ is right-continuous), we must have by Blumenthal's zero-one law that $P(\Gamma) = 1$. It follows by dominated convergence that

$$\lim_n \frac{1}{n} \sum_{k=1}^{n} P(X_{2^{-k}}(\omega) - X_{2^{-k-1}}(\omega) > 0) = \frac{1}{2}. \tag{3.3}$$

By the Brownian scaling property, we have that for $a > 0$ the processes $(X_{at})_{t\geq 0}$ and $(\sqrt{a}X_t)_{t\geq 0}$ have the same law and therefore it follows by (3.3) that

$$P(X_2 - X_1 < 0) = \frac{1}{2}.$$

This, however, is impossible since, for $\alpha < 0$,

$$
\begin{aligned}
P(X_2 - X_1 < 0) &= P\left(\alpha\left(L_2^0 - L_1^0\right) < B_1 - B_2\right) \\
&= P(B_1 - B_2 > 0) + P\left(\alpha\left(L_2^0 - L_1^0\right) < B_1 - B_2 < 0\right) \\
&> \frac{1}{2}.
\end{aligned}
$$

Similarly, $P(X_2 - X_1 < 0) < \frac{1}{2}$ for $\alpha > 0$.   □

It is possible to show via a time-change that the main result used in the proof above generalises to continuous local martingales, see Emery & Perkins (1982) for details.

Theorem 3.12. *Emery & Perkins theorem.* *Let M be a continuous local martingale, $M_0 = 0$, and L^0 be its local time at zero. For every α we have that $\mathbb{F}^M = \mathbb{F}^{M+\alpha L^0}$.*

3.4   Immediate arbitrage

Suppose that we have given a discounted price process in its canonical form

$$S = M + \int \lambda \, d\langle M \rangle, \tag{3.4}$$

and let the process K be defined as

$$K = \int \lambda^2 \, d\langle M \rangle.$$

We will discuss briefly in this section what happens if K_ε is not a.s. finite for any $\varepsilon > 0$, thus in particular the structure condition is violated.

Theorem 3.13. *Immediate arbitrage.* *Assume that S has a canonical decomposition of the form (3.4), and that for all $\varepsilon > 0$*

$$K_\varepsilon = \infty.$$

Then for all $\varepsilon > 0$, there is a strategy ϑ such that $\vartheta = \vartheta \mathbb{I}_{[0,\varepsilon]}$, $\int \vartheta \, dS \geq 0$ and for all $t > 0$

$$P\left(\int_0^t \vartheta_u \, dS_u > 0\right) = 1.$$

Proof.   See Delbaen & Schachermayer (2006), Theorem 12.3.7.   □

We shall now give an example where the conditions of Theorem 3.13 are fulfilled. With a Brownian motion B, let the price process S be given as

$$S_t = B_t - \int_0^t \frac{B_s}{s} \, ds.$$

The last integral is well defined, and moreover for all $\varepsilon > 0$

$$K_\varepsilon = \int_0^\varepsilon \frac{B_s^2}{s^2} \, ds = \infty.$$

see Jeulin & Yor (1979). Hence the conditions of Theorem 3.13 are satisfied, and accordingly, the price process allows for immediate arbitrage. This arbitrage opportunity, however, can only be realised under special circumstances. It does not exist for traders who have only information about the price process, since S is a Brownian motion with respect to its own filtration $\mathbb{F}^S$, see Chapter 1 of Yor (1992). On the other hand, arbitrage exists for well-informed traders who have access to the information contained in the larger filtration $\mathbb{F}^B$.

3.5   Super-hedging and the optional decomposition theorem

Super-hedging is a technique commonly used for the hedging of American style claims but it can also be employed to hedge European claims in incomplete markets. In the case of American options, the buyer gains the right to choose an exercise time prior to maturity T at which to receive the specified payout. Hence, the problem facing the buyer is when to exercise optimally in order to maximise the expected profit. The intuition behind super-hedging American claims is that we aim to find a portfolio that will hedge the claim regardless of whether or not it is exercised optimally. When it comes to super-hedging European claims in incomplete markets, the idea is to choose a hedging portfolio that out-performs the claim with probability one.

To make this argument more concrete, first let S be a locally bounded semi-martingale modelling the discounted price process of a stock and let $f = f(S_T)$ represent the payoff of a European claim. Also, let F be a non-negative process representing the discounted reward process of an American claim. In the case of an American call, we have $F_t = (S_t - K)^+$, but more complicated functionals of the stock price process may be used. Note that as we use discounted quantities the strike is also discounted and should read

$K = e^{-rt}K^0$ whre K^0 is the undiscounted strike. For $t \in [0, T]$ denote with $\mathcal{T}_t$ the set of all stopping times τ with values in the interval $[t, T]$. Given that the American claim has not been exercised before time t, $\mathcal{T}_t$ is the set of all the potential exercise times for the claim. To be able to approach super-hedging any claim, we firstly need to generalise the notion of value process to allow for consumption.

Definition 3.14. A **portfolio** Π is a triple (c, ϑ, C) where the constant c is the initial value of the portfolio, ϑ is an **admissible strategy** such that $\int \vartheta \, dS$ is bounded from below, and the **consumption process** C is an increasing adapted process. The **value process** V of a portfolio Π is given by

$$V = c + \int \vartheta \, dS - C.$$

If $C \equiv 0$ then the portfolio Π is called **self-financing**.

The consumption process C need not be absolutely continuous with respect to $d\langle S \rangle$, so can capture both singular 'gulps' of consumption or a continuous flow. In order to hedge the claim F we are looking for the cheapest portfolio that performs at least as well as the American claim. The idea is when the buyer exercises optimally we are perfectly hedged, whereas, if the buyer exercises sub-optimally we can liquidate the hedging portfolio to generate a cash payout. This is one reason why we allow our consumption process to be singular. The following definition makes the notion of a suitable portfolio more concrete:

Definition 3.15. A portfolio Π with value process V is called a **super-hedging portfolio** for F if $V \geq F$; the associated strategy is called the **super-hedging strategy**. A super-hedging portfolio $\widetilde{\Pi}$ with value process $\widetilde{V}$ is called **minimal** if $V \geq \widetilde{V} \geq F$ for any super-hedging portfolio Π with value process V.

The minimal initial capital required to drive a super-hedging strategy for F is

$$\overline{V}_0 := \inf \{ c \,|\, \exists \, V \text{ s.t. } V_\tau \geq F_\tau \, \forall \tau \in \mathcal{T}_0 \text{ a.s.} \}.$$

This minimum super-replicating cost is the smallest price the seller of the option is willing to accept when using super-hedging. On the other hand, the 'buyers price' is the largest amount the buyer is willing to spend on the claim and still have a possibility of making a profit, namely:

$$\underline{V}_0 = \sup \{ c \,|\, \exists \, V \text{ s.t. } V_\tau \leq F_\tau \, \forall \tau \in \mathcal{T}_0 \text{ a.s.} \}.$$

It is important to note that sometimes one has that

$$\overline{V}_0 > \underline{V}_0,$$

so there is no guarantee that the buyer and seller can agree on a price and trading might not occur.

Given the uncertainty about both the equivalent martingale measure and exercise time, the super-replicating price is defined as the maximum possible fair price.

Definition 3.16. The **super-replication price** for the American option F is given by

$$\sup_{Q\in\mathcal{M}^e} \sup_{\tau\in\mathcal{T}_0} E_Q\left[F_\tau\right]. \tag{3.5}$$

In case of a European option f maturing at T this is just

$$\sup_{Q\in\mathcal{M}^e} E_Q\left[f\right]. \tag{3.6}$$

When pricing the American claim F, we take the supremum with respect to the set of stopping times $\mathcal{T}_0$ as we should price with respect to the payoff when the buyer is behaving optimally. This is because the seller of the option expects to make his greatest loss when the buyer exercises optimally. In both cases, the supremum with respect to martingale measures is taken because of an extreme form of ambiguity aversion. It is not clear which martingale measure Q should be chosen, so the seller takes the most pessimistic view possible. In the event the seller is incorrect in his choice of measure, he can only gain from the error since he would have over-charged for the claim.

We now aim to characterise the minimal super-hedging strategy. The optional decomposition theorem allows us to characterise all value functions in terms of a potentially unknown portfolio process. This result will then be used to prove the existence of a super-replicating strategy in the sense defined above.

Theorem 3.17. *Optional decomposition theorem. Let $\mathcal{M}^e \neq \varnothing$ and consider a process V which is bounded from below and a Q-super-martingale for all $Q \in \mathcal{M}^e$. Then there exists $\vartheta \in L(S)$ and an increasing adapted process C such that*

$$V = V_0 + \int \vartheta \, dS - C, \tag{3.7}$$

i.e. V is the value process for the portfolio $\Pi = (V_0, \vartheta, C)$.

In the case when $\mathcal{M}^e$ is a singleton, a stronger result holds. In fact, the Doob-Meyer decomposition states that for any super-martingale V a decomposition (3.7) holds where C can even be chosen to be predictable, and hence is unique. Before we approach the issue of whether a super-replicating strategy exists, let us recall the following result:

Proposition 3.18. *Essential supremum*. *Let $(\Omega, \mathcal{F}, P)$ be a probability space, and $\mathcal{H}$ be a non-empty family of random variables. A random variable η is called the **essential supremum** (we write: ess sup) of $\mathcal{H}$ if it satisfies the following conditions:*

(i) for all $\xi \in \mathcal{H}$, $\xi \leq \eta$ P-a.s.
(ii) If η' is another random variable satisfying (i) then $\eta' \leq \eta$ P-a.s.

The essential supremum always exists, and when $\mathcal{H}$ contains at most a denumerable number of elements (ξ_n) then

$$\mathrm{ess\,sup}\,\mathcal{H} = \vee_n \xi_n.$$

We conclude with a result concerning the existence of the super-replicating portfolio discussed in this section.

Theorem 3.19. *Existence of a minimal super-hedging portfolio*. *Assume*

$$\sup_{Q \in \mathcal{M}^e} \sup_{\tau \in \mathcal{T}_0} E_Q\,[F_\tau] < \infty.$$

Then a minimal super-hedging portfolio exists, and its value process $\widetilde{V}$ is given by

$$\widetilde{V}_t = \mathrm{ess}\sup_{Q \in \mathcal{M}^e,\ \tau \in \mathcal{T}_t} E_Q\,[F_\tau|\,\mathcal{F}_t].$$

Proof.   (sketch): Let $Q \in \mathcal{M}^e$, $\tau \in \mathcal{T}_t$ and V be the value process of a super-hedging strategy for F. V is a local martingale minus an increasing process; since $V \geq F \geq 0$, we have that V is a Q-super-martingale. Therefore we have for any $t \in [0, T]$

$$V_t \geq E_Q\,[V_\tau|\,\mathcal{F}_t] \geq E_Q\,[F_\tau|\,\mathcal{F}_t],$$

and since Q, τ were arbitrary, $V_t \geq \widetilde{V}_t$. Moreover, we have since $t \in \mathcal{T}_t$ that

$$\widetilde{V}_t = \mathrm{ess}\sup_{Q \in \mathcal{M}^e,\ \tau \in \mathcal{T}_t} E_Q\,[F_\tau|\,\mathcal{F}_t] \geq \mathrm{ess}\sup_{Q \in \mathcal{M}^e} E_Q\,[F_t|\,\mathcal{F}_t] = F_t.$$

It remains to show that there exists indeed a portfolio $\widetilde{\Pi}$ which has $\widetilde{V}$ as value process. The key observation (which we do not prove here; see

Föllmer & Kramkov (1997)) is that $\widetilde{V}$ is a Q-super-martingale with respect to all $Q \in \mathcal{M}^e$. The result then follows from the optional decomposition theorem. □

3.6 Arbitrage via a non-equivalent measure change

It is sometimes useful to construct asset price processes using a non-equivalent measure change as this creates the opportunity to place bounds on the asset price. To do this we first specify unconstrained dynamics under a martingale measure P^0 and then change to a measure Q that is absolutely continuous but not equivalent to P^0 in order to constrain the price process. However, such a procedure has the potential to induce arbitrage opportunities. The purpose of this section is to illustrate how the optional decomposition theorem can be applied to prove that, under certain conditions, there exist arbitrage opportunities under the non-equivalent measure Q.

To further motivate this discussion we shall first discuss an example in a complete FX-market where arbitrage opportunities are created once the domestic currency is tied to some foreign currency by allowing it to float freely only within a certain range. As usual, we work on a filtered probability space $(\Omega, \mathcal{F}, \mathbb{F}, P^0)$ with a fixed time horizon $T > 0$. The exchange rate process S that is used to convert foreign payoffs into domestic currency, is modelled under P^0 for simplicity as

$$\frac{dS}{S} = \sigma \, dB,$$

where σ is some positive constant and B is a P^0-Brownian motion. We assume that regulation restricts S to move only in a range of $[a, b]$ for some $b > a > 0$; that the regulatory authority is capable of supporting the currency in this manner and also that $S_0 \in (a, b)$.

Observe that we have for the unrestricted dynamics

$$P^0 \left(a \leqslant S_t \leqslant b \quad \forall t \in [0, T] \right) > 0$$

and

$$P^0 \left(\exists t \in [0, T] \text{ s.t. } S_t \notin [a, b] \right) > 0. \tag{3.8}$$

The regulatory impact is reflected in the new measure Q, which is defined via its density:

$$Z = \frac{dQ}{dP^0} = \begin{cases} 0 & \text{if } S_t \notin [a, b] \text{ for some } t \in [0, T] \\ c & \text{otherwise} \end{cases},$$

where c is a normalising constant. We consider the contingent claim

$$f = \mathbb{I}_{Z>0}.$$

As the original market under P^0 is complete, the value process of the claim, $V_t = E_{P^0}\left[f\,|\,\mathcal{F}_t\right] > 0$, has the PRP with an admissible strategy ϑ since we have that

$$\int_0^t \vartheta_u\, dS_u = V_t - V_0 \geqslant -V_0 = -E_{P^0}(f) \geqslant -1 \qquad P^0 \text{ and } Q - \text{a.s.}$$

An arbitrage opportunity exists under Q, since

$$\int_0^T \vartheta_t\, dS_t = f - V_0 = 1 - E_{P^0}(f) > 0 \qquad Q - \text{a.s.}$$

This is because $Q(Z = 0) = 0$ implies that $f = 1$, Q–a.s. while on the other hand, (3.8) implies that $E_{P^0}(f) < 1$.

This example motivates the next proposition which proves using the optional decomposition theorem that under a certain condition an absolutely continuous but non-equivalent measure change can create arbitrage opportunities in incomplete markets.

Proposition 3.20. Arbitrage via a non-equivalent change of measure. *Let S be a P^0-local martingale, and consider a probability measure Q which is absolutely continuous, but not equivalent to P^0. If*

$$\sup_{P \in \mathcal{M}^e} P\left(\frac{dQ}{dP^0} > 0\right) < 1 \tag{3.9}$$

then there exists an arbitrage opportunity under Q which can be realised via an admissible strategy, i.e. a strategy ϑ such that $\int \vartheta\, dS$ is bounded from below.

Proof. We consider the claim

$$f := \mathbb{I}_{\frac{dQ}{dP^0} > 0}$$

and define the process V as

$$V_t = \text{ess} \sup_{P \in \mathcal{M}^e} E_P\left[f\,|\mathcal{F}_t\right], \qquad t \geqslant 0.$$

V is a P-super-martingale for all $P \in \mathcal{M}^e$, see Föllmer & Kramkov (1997). It follows (since $P^0 \in \mathcal{M}^e$) that for $t \geqslant 0$,

$$V_t \geqslant E_{P^0}\left[f\,|\mathcal{F}_t\right] \geqslant 0 \qquad P^0 - \text{a.s.}$$

By the optional decomposition theorem, there exist $\vartheta \in L(S)$ and an increasing adapted process C such that

$$V = V_0 + \int \vartheta \, dS - C.$$

Moreover, ϑ is admissible since $\int \vartheta \, dS$ is bounded from below: for $t \geqslant 0$,

$$\int_0^t \vartheta_u \, dS_u = V_t - V_0 + C_t \geqslant -V_0$$

$$= - \sup_{P \in \mathcal{M}^e} E_P \left[f \right] \geqslant -1 \qquad P^0 \text{ and } Q - \text{a.s.}$$

For ϑ to be an arbitrage opportunity under Q, we need to check whether

$$\int_0^T \vartheta_t \, dS_t = f - V_0 + C_T > 0 \qquad Q - \text{a.s.}$$

Since we have that $f = 1$ Q-a.s., $C_0 = 0$ and C is increasing, this holds in particular if $1 - V_0 > 0$ or

$$\sup_{P \in \mathcal{M}^e} E_P \left[f \right] < 1,$$

which is equivalent to

$$\sup_{P \in \mathcal{M}^e} P \left(\frac{dQ}{dP^0} > 0 \right) < 1,$$

which is our assumption (3.9). $\qquad \square$

Remark. The filtration $\mathbb{F}$ in this section typically does not satisfy the usual conditions with respect to Q. However, there is a remedy: consider the filtration $\mathbb{G}$ obtained from $\mathbb{F}$ by adding all Q-null sets. The results in Delbaen & Schachermayer (1995b) then show that whenever we have a stopping time τ_Q and a $\mathbb{G}$-predictable process ϑ_Q there exists a stopping time τ_P and an $\mathbb{F}$-predictable process ϑ_P such that Q-a.s. $\tau_Q = \tau_P$ and ϑ_Q and ϑ_P are Q-indistinguishable. This implies that whenever we need to work with a $\mathbb{G}$-predictable process we can essentially replace it by an $\mathbb{F}$-predictable process.

At this point, once we have made an absolutely continuous but non-equivalent measure change, it is sensible to question how can we describe the price dynamics under the measure Q. In the case of continuous price processes we can deploy Lenglart's extension of Girsanov's theorem; see Protter (2005), Corollary to Theorem III.42.

Theorem 3.21. Lenglart's theorem. *Let Q be a probability measure absolutely continuous with respect to P^0. Define the process Z as*

$$Z_t = E_{P^0}\left[\frac{dQ}{dP^0}\,\Big|\,\mathcal{F}_t\right].$$

Let S be a continuous local martingale under P^0. Then there exists an S-integrable process α such that

$$S - \int \frac{1}{Z_-}\,d\,[Z,S] = S - \int \alpha\,d\,[S]$$

is a Q-local martingale.

In general, it is difficult to find an explicit expression for the drift α, but the following approach can be useful. As the process Z in Lenglart's theorem is square-integrable, we can use the Kunita-Watanabe decomposition to project Z on the space of all square-integrable martingales that can be written as $\int \gamma\,dS$ for some $\gamma \in L^2(S)$. The resulting orthogonal projection, denoted $\int \beta\,dS$, is square-integrable by construction and such that $[Z, S] = \int \beta\,d\,[S]$. The process Z is Q-a.s. strictly positive, so we can set

$$\alpha = \frac{\beta}{Z_-}.$$

It results that

$$\int \frac{1}{Z_-}\,d\,[Z,S] = \int \frac{\beta}{Z_-}\,d\,[S]$$
$$= \int \alpha\,d\,[S].$$

Summing up, the dynamics of S under Q is given as

$$dS = d\widetilde{S} + \alpha\,d\,[S], \tag{3.10}$$

where $\widetilde{S} = S - \int \alpha\,d\,[S]$ is a local Q-martingale.

To conclude we shall briefly discuss the problem of valuing claims in the setting discussed in this section. It is not appropriate to define a price based on a super-replication argument and using admissible integrands, as this would lead to a non-finite price. Fortunately, it turns out that pricing is still possible if we only allow strategies which require no intermediate credit. To price an integrable claim f we need to consider the modified claim $\widetilde{f} = f\,\mathbb{I}_{\Omega_Q}$ where Ω_Q is the support of the measure Q. If we assume that the market under the martingale measure P^0 is complete then we can assign to $\widetilde{f}$ the usual no-arbitrage price $E_{P^0}[\widetilde{f}]$. Note that if $f \geqslant 0\ Q$−a.s, the associated value process satisfies

$$V_t = E_{P^0}\left[f\,\mathbb{I}_{\Omega_Q}\,\Big|\,\mathcal{F}_t\right] \geqslant 0 \qquad Q - \text{a.s.}$$

3.7 Notes and further reading

A very good and accessible introduction to mathematical finance is Protter (2001). Arbitrage theory on an advanced level is treated in Delbaen & Schachermayer (2006), and general formulations of the two fundamental theorems of asset pricing can be found in Cerny & Shiryaev (2002). Complete markets with discontinuous security prices have been studied by Dritschel & Protter (1999). The section about hidden arbitrage is based on Jarrow & Protter (2005) and Emery & Perkins (1982). Immediate arbitrage has been introduced in Delbaen & Schachermayer (1995a), our example is due to Jeulin & Yor (1979). The optional decomposition theorem in the form stated here is due to Föllmer & Kramkov (1997).

In case of a European claim, it turns out by (3.6) that the super-replication price is too expensive. Therefore, it has been proposed in Föllmer & Leukert (1999) to maximise the probability of a successful super-hedge given a certain amount of initial capital, a concept which they call *quantile hedging*. However, with this approach there is no protection for the worst case scenarios other than portfolio diversification, and technically it might be difficult to implement since it corresponds to hedging a knock-out option. Föllmer & Leukert (2000) moreover considered *efficient hedges* which minimise the expected shortfall weighted by some loss function. In this way the investor may interpolate between the extremes of no hedge and a super-hedge, depending on the accepted level of shortfall risk.

The section about arbitrage by a non-equivalent measure change is based on Osterrieder & Rheinländer (2006) who were motivated by the study of arbitrage opportunities in diverse markets by Fernholz, Karatzas & Kardaras (2005). The description of the drift in Lenglart's theorem comes from Jacod (1979). An early study about arbitrage via a non-equivalent measure change is Gossen-Dombrowsky (1992) who considers a stock price that follows geometric Brownian motion respecting two a priori fixed exponential curves as upper and lower boundaries. Also, Delbaen & Schachermayer (1995b) use a Brownian motion B with $B_0 = 1$ and construct a new measure which assigns probability zero to the set of paths where B ever hits zero in order to show that 3-dimensional Bessel processes cause arbitrage opportunities.

Chapter 4

Asset Price Models

In this chapter we present our main two example classes of models for financial assets. These are exponential Lévy processes and continuous stochastic volatility models. The former class may exhibit jumps, and the distribution of a Lévy process $X = (X_t)$ is already determined by the distribution of the random variable X_1. The latter class, in contrast, displays a wide range of truly dynamic phenomena like the occurrence of financial bubbles.

4.1 Exponential Lévy processes

4.1.1 *A Lévy process primer*

Definition 4.1. A process X defined on $(\Omega, \mathcal{F}, P)$ is said to be a **Lévy process** if for $0 \leq s \leq t$ it possesses the following properties:

(i) The paths of X are almost surely right continuous with left limits;
(ii) X has stationary increments, i.e. $X_t - X_s$ is equal in distribution to X_{t-s};
(iii) The increments are independent of the past: $X_t - X_s$ is independent of $\sigma(X_u : u \leq s)$.

The filtration $\mathbb{F}^X$ generated by a Lévy process X has benign properties, see Protter (2005), Theorem I.31. Note that by our convention, $\mathbb{F}^X$ is the augmented natural filtration.

Proposition 4.2. *Lévy filtration*. *Let X be a Lévy process, then $\mathbb{F}^X$ fulfills the usual conditions of completeness and right-continuity.*

The next result introduces the key object describing the jump behaviour of a Lévy process.

Proposition 4.3. Lévy measure. *If X is a Lévy process, define a set function ν on $\mathbb{R}$ by setting for every Borel set Γ which does not contain zero*

$$\nu(\Gamma) := E\left[\sum_{0 < t \leq 1} \mathbb{I}_\Gamma(\Delta X_t)\right], \qquad (4.1)$$

*and $\nu(0) := 0$. Then ν is a positive measure, called the **Lévy measure**, which has the properties:*

(i) $\nu(\Gamma) < \infty$ if Γ is a Borel set and bounded away from zero;
(ii) $\int_{|x| \leq 1} x^2\, \nu(dx) < \infty$.

By its definition (4.1), $\nu(\Gamma)$ is the expected number of jumps per unit time whose sizes fall into the set Γ.

We model the price process S of a risky asset as the exponential of a Lévy process X with $X_0 = 0$,

$$S = S_0 \exp(X).$$

To get a reasonable model, which in particular does not allow for arbitrage opportunities, we will assume that there exists an equivalent martingale measure for S. Consequently, by (2.7), S is a *special* semi-martingale.

Definition 4.4. A semi-martingale X is called **exponentially special** if $\exp(X)$ is a special semi-martingale.

While under the usual assumption of existence of an equivalent martingale measure for $S = S_0 \exp(X)$ we have that X is exponentially special, it is in many cases of interest whether the log-return process X itself is a special semi-martingale. The following criterion is from Kallsen & Shiryaev (2002), Lemma 2.8.

Lemma 4.5. *A Lévy process X is a special semi-martingale if and only if*

$$\int_{|x|>1} |x|\, \nu(dx) < \infty.$$

Thus the large jump behaviour determines whether X is special, but it is the distribution of the small jumps which yields important information about the paths of X, see Sato (1999), Theorem 21.9.

Lemma 4.6. *A pure jump Lévy process X is of finite variation if and only if*

$$\int_{|x| \leq 1} |x|\, \nu(dx) < \infty.$$

In our study of exponential Lévy processes we sometimes need to assume the existence of certain exponential moments.

Condition (EM). *There is a constant $\beta > 0$ such that for every $u \in [-\beta, \beta]$ we have*

$$\int_{|x|>1} e^{ux} \, \nu(dx) < \infty.$$

Note that (EM) in particular implies that X is a special semi-martingale. A consequence of condition (EM) is the existence of certain exponential moments, see Sato (1999), Theorem 25.3.

Proposition 4.7. *Exponential moments of Lévy processes.* (EM) *holds if and only if*

$$E\left[e^{uX_t}\right] < \infty \quad \text{for every } t > 0,$$

and for all $u \in [-\beta, \beta]$.

Definition 4.8. Let X be a Lévy process. By setting

$$D := \left\{ z \in \mathbb{C} \,\middle|\, E\left[e^{\Re(z)\,X_1}\right] < \infty \right\},$$

we define for $t > 0$ the **cumulant function** κ as

$$\kappa : D \to \mathbb{C}$$
$$e^{\kappa(z)t} = E\left[e^{zX_t}\right].$$

The cumulant function exists at least for $z \in \mathbb{C}$ with $\Re(z) = 0$, and in this case $\kappa(iu)$ coincides with the **characteristic exponent**

$$\psi(u) := \log E\left[e^{iuX_1}\right], \quad u \in \mathbb{R},$$

which often has a simpler form than the distribution of X_1 which it determines uniquely. The cumulant function can be used to compensate exponentials of Lévy-processes as follows.

Proposition 4.9. *Exponential martingale.* *Let X be a Lévy process with cumulant function κ. For any $\vartheta \in \mathbb{R}$ such that $E\left[\exp(\vartheta X_1)\right] < \infty$, we have that $M := \exp(\vartheta X - \kappa(\vartheta)\,t)$ is a martingale with mean one.*

Proof. Adaptedness to the filtration $\mathbb{F} = (\mathcal{F}_t)$ generated by X is obvious since $\kappa(\vartheta)$ is deterministic, and integrability follows by assumption. Moreover, by stationary independent increments and the definition of $\kappa(\vartheta)$ we have for $t > s$

$$E\left[M_t \mid \mathcal{F}_s\right] = M_s \, E\left[e^{\vartheta(X_t - X_s)} \mid \mathcal{F}_s\right] e^{-\kappa(\vartheta)(t-s)}$$
$$= M_s \, E\left[e^{\vartheta X_{t-s}}\right] e^{-\kappa(\vartheta)(t-s)}$$
$$= M_s.$$

$\square$

Definition 4.10. Let $\vartheta \in \mathbb{R}$ such that $E\left[\exp(\vartheta X_1)\right] < \infty$. The probability measure $Q^\vartheta \sim P$, defined via the density process

$$\left.\frac{dQ^\vartheta}{dP}\right|_{\mathcal{F}_t} = e^{\vartheta X_t - \kappa(\vartheta)t},$$

is called the **Esscher transform** of P (with respect to ϑ).

For further study of the cumulant function we will need the following concept which allows us to deal with the small jumps separately.

Definition 4.11. A **truncation function** is a bounded measurable function h such that $h(x) = x$ in a neighbourhood of zero. A common choice is $h(x) = x\mathbb{I}_{|x|\leq 1}$.

There are two important cases where simple choices of h are possible. We will refer to them, by a slight abuse of terminology, as truncation functions as well. When X is a special semi-martingale we may choose the identity $h(x) = x$; whereas if the jump part of X is of finite variation we may choose $h(x) = 0$.

We have the following general form of the cumulant function relative to a fixed truncation function h.

Theorem 4.12. Lévy-Khintchine formula. *Let X be a Lévy process. Then there exists a triplet (γ, σ^2, ν) such that the cumulant function can be written for $z \in D$ as*

$$\kappa(z) = \gamma z + \frac{1}{2}\sigma^2 z^2 + \int \left(e^{zx} - 1 - zh(x)\right)\nu(dx),$$

*where γ and σ^2 are constants, and ν is the Lévy measure. (γ, σ^2, ν) is called the **characteristic triplet** of the Lévy process X. Neither σ^2 nor ν depend on the choice of truncation function h, but γ does.*

Let us give two examples of Lévy processes.

(A) Brownian motion

A **Brownian motion** X with drift μ and volatility $\sigma > 0$ is a Lévy process with a.s. continuous paths such that the increments $X_t - X_s$, $t > s$, are normally distributed with mean $\mu(t-s)$ and variance $\sigma^2(t-s)$. Its characteristic exponent is

$$\psi(u) = \mu i u - \frac{1}{2}\sigma^2 u^2,$$

whereas, the path continuity implies that the Lévy measure is the zero measure. Comparing with the Lévy-Khintchine formula, we have the correspondences $\gamma = \mu$, $\sigma^2 = \sigma^2$, and $\nu = 0$, hence (EM) is trivially fulfilled. As for exponential moments, we have that for any real u,

$$E\left[e^{uX_t}\right] = \exp\left(\mu u t + \frac{\sigma^2 u^2 t}{2}\right).$$

If $X_0 = 0$, $\mu = 0$ and $\sigma = 1$ then X is called **standard Brownian motion**, and usually denoted by B.

(B) Compound Poisson process

Let N be a Poisson process with intensity $\lambda > 0$ and (Y_i) a sequence of i.i.d. random variables, with common law F, and which are independent of N. The process X given as

$$X_t = \sum_{i=1}^{N_t} Y_i, \qquad t \geq 0,$$

is called **compound Poisson process**. The expected number of jumps per time unit of X is λ, and the jump size distribution is given by the distribution F of Y_i. By (4.1), this determines the Lévy-measure as

$$\nu(\Gamma) = \lambda P(Y_i \in \Gamma).$$

Furthermore, taking $h(x) = 0$ since X is of finite variation, the Lévy-Khintchine formula gives that the characteristic exponent is

$$\psi(u) = \int \left(e^{iux} - 1\right) \nu(dx)$$

$$= \lambda \int \left(e^{iux} - 1\right) F(dx).$$

Comparing with the Lévy-Khintchine formula, we get $\gamma = 0$, $\sigma^2 = 0$, and $\nu(dx) = \lambda F(dx)$. For (EM) to be satisfied, we have to assume that F possesses certain exponential moments.

A Lévy process can be of one of three different types, see Sato (1999):

- **Type A:** when $\sigma^2 = 0$ and $\nu(\mathbb{R}) < \infty$. X is a compound Poisson process, in particular has finite variation, and the time of the first jump is exponentially distributed with mean $1/\nu(\mathbb{R})$.
- **Type B:** when $\sigma^2 = 0$, $\nu(\mathbb{R}) = \infty$, and $\int_{|x| \leq 1} |x| \nu(dx) < \infty$. X is a finite variation process, but its jumping times are dense in $\mathbb{R}_+$, that is, they are not enumerable in increasing order.
- **Type C:** when either $\sigma^2 > 0$ or $\int_{|x| \leq 1} |x| \nu(dx) = \infty$. X is a process of infinite variation. In the case when $\sigma^2 > 0$ it has a Brownian component. For type C processes, the sum over small jumps might not converge, and therefore has to be compensated.

For any adapted cadlag process X one can define an integer-valued random measure μ^X, called the **jump measure**, by setting

$$\mu^X(\omega; dt, dx) = \sum_s \mathbb{I}_{\{\Delta X_s(\omega) \neq 0\}} \delta_{(s, \Delta X_s(\omega))} (dt, dx).$$

Here δ denotes the Dirac measure. Intuitively, μ^X is a counter which increases whenever within a time increment dt a jump occurs whose size falls into dx.

For any Borel function f such that the subsequent integral exists, we can now define

$$\left(f(x) * \mu^X\right)_t := \int_0^t \int f(x) \, \mu^X(ds, dx)$$

$$= \sum_{0 < s \leq t} f(\Delta X_s).$$

The following two results are from Jacod & Shiryaev (2003), Theorem II.1.8 and Proposition II.1.28, respectively.

Proposition 4.13. *Predictable compensator.* *If* $|f(x)| * \mu^X$ *is of locally integrable variation, then the measure* $\nu^X = \nu \otimes dt$, *called the* **predictable compensator** *of* μ^X, *is such that* $f(x) * \left(\mu^X - \nu^X\right)$ *is a local martingale.*

Lemma 4.14. *If* $|f(x)| * \mu^X$ *(or, equivalently* $|f(x)| * \nu^X$*) is of locally integrable variation, we can write*

$$f(x) * \left(\mu^X - \nu^X\right) = f(x) * \mu^X - f(x) * \nu^X.$$

Let us now state one of the main results about Lévy processes, see He *et al.* (1992), Theorem 11.45.

Theorem 4.15. Lévy-Itô decomposition. *We can represent any Lévy process X relative to a truncation function h as*

$$X = \gamma t + \sigma B + h(x) * \left(\mu^X - \nu^X\right) + (x - h(x)) * \mu^X. \tag{4.2}$$

Here γ and $\sigma > 0$ are constants, and B is a standard Brownian motion. Moreover, if X is a special semi-martingale, we can write

$$X = \gamma t + \sigma B + x * \left(\mu^X - \nu^X\right), \tag{4.3}$$

and in the case when the jump part of X is of finite variation we may write

$$X = \gamma t + \sigma B + x * \mu^X. \tag{4.4}$$

Note that γ depends on the choice of the truncation function h.

In the Lévy framework, there are no problems caused by stochastic exponentials being strict local martingales.

Theorem 4.16. Lévy martingales. *A local martingale M which is a Lévy process is a martingale. Moreover, if $\Delta M > -1$ then its stochastic exponential $\mathcal{E}(M)$ is a positive martingale.*

Proof. See He *et al.* (1992), Theorem 11.46, for the first statement, and Protter & Shimbo (2008), Corollary 7, for the second. $\square$

4.1.2 *Examples of Lévy processes*

A **subordinator** Z is a non-decreasing Lévy process, hence it can be thought of as a random time-change. The Lévy measure of a subordinator is concentrated on the positive real axis. More precisely, it is shown in Applebaum (2009), Theorem 1.3.15, that a Lévy process is a subordinator if and only if its cumulant function can be written as

$$\kappa(z) = \gamma z + \int_0^\infty (e^{zx} - 1)\, \nu(dx), \tag{4.5}$$

with $\gamma \geq 0$, and the Lévy measure ν satisfies

$$\nu(-\infty, 0) = 0, \qquad \int_0^\infty (x \wedge 1)\, \nu(dx) < \infty.$$

Next we provide two examples of common subordinators.

(A) Gamma process (Type B)

Recall that the density of the gamma distribution with parameters $a > 0$ and $b > 0$ is given by

$$f_\Gamma(x; a, b) = \frac{b^a}{\Gamma(a)} x^{a-1} e^{-bx}, \qquad x > 0.$$

The characteristic exponent has the simple form

$$\psi_\Gamma(u; a, b) = a \log\left(1 + \frac{u}{b}\right).$$

We have that the gamma(a, b)-distribution has a kurtosis of $3\left(1 + 2a^{-1}\right)$, hence is leptokurtic, i.e. has heavier tails than the normal distribution.

The **gamma process** X is now defined as the stochastic process which starts at zero and has stationary and independent gamma(a, b)-distributed increments, so X_t is gamma(at, b)-distributed. An analytic calculation reveals that the characteristic exponent has the form (4.5) with $\gamma = 0$ and Lévy measure

$$\nu_\Gamma(dx) = \frac{a}{x} e^{-bx}\, dx$$

for $x > 0$ and zero elsewhere. Therefore, the gamma process is a subordinator.

(B) Inverse Gaussian process (Type B)

For $b > 0$, consider Brownian motion with drift $X = B + bt$ and let

$$T_t = \inf\{s > 0\,|\,X_s = at\}$$

be the first hitting time of level $at > 0$. The mapping $t \mapsto T_t$ is a.s. non-decreasing, and is the generalised inverse of a Gaussian process. It is called the **inverse Gaussian subordinator**. The density function of T_1 which can be calculated using the reflection principle is

$$f_{IG}(x; a, b) = \frac{a}{\sqrt{2\pi}} \exp(ab)\, x^{-3/2} \exp\left(-\frac{1}{2}\left(a^2 x^{-1} + b^2 x\right)\right), \qquad x > 0.$$

The characteristic exponent is given as

$$\psi_{IG}(u; a, b) = -a\left(\sqrt{-2iu + b^2} - b\right).$$

We define the **inverse Gaussian process** X as the stochastic process which starts at zero and has stationary and independent inverse Gaussian distributed increments. Its Lévy measure is given as

$$\nu_{IG}(dx) = \frac{a}{\sqrt{2\pi}} x^{-3/2} \exp\left(-\frac{1}{2} b^2 x\right) dx$$

for $x > 0$ and zero elsewhere.

4.1.3 *Construction of Lévy processes by subordination*

The arrival process of market or limit orders for financial assets is not homogenous, but instead clustered in time. Therefore, we may distinguish between real time and business time where the latter is proportional to the order arrival process. It is a natural idea to construct asset price models by starting with a Lévy process X which we subsequently time-change using an independent subordinator T which models business time. The resulting process Z given by $Z_t := X_{T_t}$ is again a Lévy process, see Applebaum (2009), Theorem 1.3.25.

(A) Variance gamma process (Type B)

Let $\sigma > 0$, $\mu \in \mathbb{R}$, and take a gamma process G with parameters $a = b = 1/v$, for $v > 0$. The **VG process** Z^{VG} is now constructed by gamma-subordinating Brownian motion with drift,

$$Z_t^{VG} = \mu G_t + \sigma B_{G_t}.$$

Alternatively, one can also construct Z^{VG} as difference of two independent gamma processes, from which the Type B property follows. When $\mu = 0$, the VG-distribution is symmetric with variance σ^2 and kurtosis $3(1 + \nu)$. The VG process is characterised by the three parameters σ, ν, and μ. It satisfies condition (EM), and its characteristic exponent is

$$\psi_{VG}(u) = -\frac{1}{\nu} \log \left(1 - i\mu\nu u + \frac{1}{2}\sigma^2 \nu u^2 \right).$$

An extension of the VG process is provided by the $CGMY$ process which contains an additional parameter determining the path behaviour; see Carr *et al.* (2002), (2003). The Lévy measure of the $CGMY$ process is given as

$$\nu(dx)/dx = \frac{C}{x^{Y+1}} e^{-Mx} \mathbb{I}_{x>0} + \frac{C}{|x|^{Y+1}} e^{G|x|} \mathbb{I}_{x<0},$$

where $C > 0$, $M \geq 0$, $G \geq 0$ and $Y < 2$. If $Y < 0$, there are finitely many jumps in any finite interval (Type A); for $Y \in [0, 1)$ the process has infinite activity, but finite variation (Type B); whereas for $Y \geq 1$ the $CGMY$ process is of infinite variation (Type C). The VG process corresponds to the choices

$$C = 1/\nu, \ G = 2\left(-\mu\nu + \sqrt{\mu^2\nu^2 + 2\nu\sigma^2} \right)^{-1},$$

$$M = 2\left(\mu\nu + \sqrt{\mu^2\nu^2 + 2\nu\sigma^2} \right)^{-1}, \ Y = 0.$$

(B) Normal inverse Gaussian process (Type C)

Take four parameters $\alpha > 0$, $\beta \in [-\alpha, +\alpha]$, $\delta > 0$ and $\mu \in \mathbb{R}$. The NIG process Z^{NIG} is constructed by time-changing Brownian motion with drift μ via an inverse Gaussian process I with parameters $a = 1$ and $b = \delta \sqrt{\alpha^2 - \beta^2}$,

$$Z_t^{NIG} = \beta \delta^2 I_t + \delta B_{I_t}.$$

When $\beta = 0$, the distribution of Z_1^{NIG} is symmetric with variance δ/α and kurtosis $3(1 + 1/\delta\alpha)$. The exponential moments $E\left[\exp\left(uZ_1^{NIG}\right)\right]$ are finite whenever $|\beta + u| < \alpha$, and therefore the NIG process satisfies condition (EM). Its cumulant function is

$$\kappa_{NIG}(z) = \mu z + \delta \left(\sqrt{\alpha^2 - \beta^2} - \sqrt{\alpha^2 - (\beta + z)^2} \right),$$

which exists for z satisfying

$$-\alpha - \beta \leq \Re(z) \leq \alpha - \beta.$$

The Lévy measure is

$$\nu(dx) = \frac{\alpha \delta}{\pi} e^{\beta x} |x|^{-1} K_1(\alpha |x|)\, dx,$$

K_1 denoting the modified Bessel function of the third kind with index 1.

(C) Hyperbolic process (Type C)

This is a four-parameter process, characterised by its cumulant function which equals

$$\kappa_H(z) = \mathrm{Ln} \left(\frac{\sqrt{\alpha^2 - \beta^2}}{\sqrt{\alpha^2 - (\beta + z)^2}} \frac{K_1\left(\delta\sqrt{\alpha^2 - (\beta + z)^2}\right)}{K_1\left(\delta\sqrt{\alpha^2 - \beta^2}\right)} e^{\mu z} \right)$$

and exists for z such that

$$-\alpha - \beta \leq \Re(z) \leq \alpha - \beta.$$

Here Ln denotes the distinguished logarithm (see Sato (1999), Lemma 7.6). For the NIG process as well as the hyperbolic process, α is measuring the steepness of the distribution; β the skewness (with $\beta = 0$ corresponding to the symmetric case); δ is a scaling and μ a location parameter. Both the NIG and hyperbolic processes are pure jump processes, so have no Brownian component. These distributions are both members of the family of **generalised hyperbolic** distributions which also contains the VG-distribution as limiting case, for details see Eberlein & Raible (2001).

4.1.4 *Risk-neutral Lévy modelling*

It is often convenient to directly model the price process S under some martingale measure Q. As we are interested in option pricing and hedging, it is the Q-measure which ultimately matters and not the real world measure P. Although exponential Lévy models are in general incomplete, we take the point of view that option prices are obtained by taking the expectation under a martingale measure chosen by the market. Risk neutral parameters of the process can then be found by calibration to a set of plain vanilla options.

Hence, in principle, the price process must be chosen to be a (local) martingale. This can be directly achieved by taking S as the stochastic exponential of a Lévy process Y which is a martingale. The local martingale property follows since $S = S_0 \, \mathcal{E}(Y)$ satisfies the stochastic integral equation

$$S = S_0 + \int S_- \, dY.$$

By Theorem 4.16, S is then even a true martingale.

Sometimes, for example in the context of the Laplace approach to option pricing, it is more convenient to model S as ordinary exponential of a Lévy process, $S = S_0 \exp(X)$. However, it turns out that these two representations can be expressed in terms of each other.

Proposition 4.17. *Equivalence between two different representations of exponential Lévy processes.* *Let* X, Y *be two Lévy processes such that* $X_0 = 0$ *and* $\Delta Y > -1$. *Then we have* $\exp(X) = \mathcal{E}(Y)$ *if and only if*

$$X = Y - Y_0 - \frac{1}{2}[Y^c] + (\log(1 + y) - y) * \mu^Y,$$

$$Y = Y_0 + X + \frac{1}{2}[X^c] + (e^x - 1 - x) * \mu^X,$$

$$x * \mu^X = \log(1 + y) * \mu^Y, \qquad y * \mu^Y = (e^x - 1) * \mu^X.$$

Proof. As Y is a Lévy process with $\Delta Y > -1$, the stochastic exponential $Z = \mathcal{E}(Y)$ is well-defined and satisfies

$$dZ = Z_- \, dY, \qquad \frac{Z}{Z_-} = \Delta\left((1 + y) * \mu^Y\right).$$

Hence $X = \log Z$ is given by Itô's formula as

$$X = \int \frac{dZ}{Z_-} - \frac{1}{2} \int \frac{d\,[Z^c]}{Z_-^2} + \sum \left(\log \left(\frac{Z}{Z_-} \right) - \frac{Z - Z_-}{Z_-} \right)$$

$$= Y - Y_0 - \frac{1}{2}\,[Y^c] + (\log(1 + y) - y) * \mu^Y.$$

This yields immediately the formulae about the jumps, as well as $[Y^c] = [X^c]$. It follows that

$$Y = Y_0 + X + \frac{1}{2}\,[Y^c] - (\log(1 + y) - y) * \mu^Y$$

$$= Y_0 + X + \frac{1}{2}\,[X^c] + (e^x - 1 - x) * \mu^X.$$ □

Corollary 4.18. *Martingale condition.* *Let X be a Lévy process such that $X_0 = 0$, condition (EM) holds with $\beta = 1$, and with Lévy-Itô decomposition (we may choose the identity as truncation function)*

$$X = \gamma t + \sigma B + x * \left(\mu^X - \nu^X \right).$$

Then we have that $S = S_0 \exp(X)$ is a martingale if and only if

$$\gamma + \frac{\sigma^2}{2} + \int (e^x - 1 - x)\, \nu\,(dx) = 0. \tag{4.6}$$

In which case, the dynamics of S can be written as

$$\frac{dS}{S_-} = \sigma\, dB + d\,(e^x - 1) * \left(\mu^X - \nu^X \right). \tag{4.7}$$

Proof. According to the previous proposition, we have

$$\int \frac{dS}{S_-} = X + \frac{\sigma^2}{2}t + (e^x - 1 - x) * \mu^X$$

$$= \left(\gamma + \frac{\sigma^2}{2} \right) t + \sigma B + (e^x - 1) * \left(\mu^X - \nu^X \right)$$

$$+ (e^x - 1 - x) * \nu^X$$

$$= M + A$$

where the local martingale part M is given as

$$M = \sigma B + (e^x - 1) * \left(\mu^X - \nu^X \right),$$

which yields (4.7). The finite variation part A is

$$A = \left(\gamma + \frac{\sigma^2}{2} + \int (e^x - 1 - x)\, \nu\,(dx) \right) t.$$

It follows that S is a martingale if and only if (4.6) holds. □

4.1.5 *Weak representation property and measure changes*

Assume that the filtration $\mathbb{F}$ is generated by a Lévy process Y with jump measure and compensator denoted by μ^Y and $\nu^Y(dx, dt) = \nu(dx)dt$ respectively. The continuous martingale part of Y is (modulo a multiplicative constant) a Brownian motion denoted B. We denote by $L^2(\mu^Y)$ the space of predictable functions $\psi = \psi(t, \omega, x)$ such that

$$\int_0^t \int_{|x|\leq 1} |\psi(s, x)| \; \nu(dx) \; ds < \infty,$$

$$\int_0^t \int_{|x|>1} \psi^2(s, x) \; \nu(dx) \; ds < \infty.$$

For the following result see Jeanblanc *et al.* (2009), Proposition 11.2.8.1.

Theorem 4.19. *Weak representation property.* *Every local $(\mathbb{F}, P)$-martingale M with $M_0 = 0$ can be written as*

$$M = \int \sigma^M dB + W^M(x) * (\mu^Y - \nu^Y)$$

where $\sigma^M \in L^2(B)$ and $W^M \in L^2(\mu^Y)$.

We now turn our attention to **structure preserving** measure changes for a Lévy process X. These are given by equivalent probability measures Q such that X remains a Lévy process under Q. The fundamental result about such measure changes states that these are described by two quantities β, y where β is a real number and y a deterministic, time-independent function.

Definition 4.20. A pair (β, y) are called **Girsanov parameters** when $\beta \in \mathbb{R}$ and $y : \mathbb{R} \to \mathbb{R}_+$ is a Borel measurable function such that

$$\int \left(\sqrt{y(x)} - 1 \right)^2 \nu(dx) < \infty.$$

Theorem 4.21. *Structure preserving measure transforms.* *Let X be a Lévy process with characteristic triplet (γ, σ^2, ν) relative to P and the truncation function h. Then a probability measure $Q \sim P$ is structure preserving if and only if there exist Girsanov parameters (β, y) such that the characteristics $(\gamma', (\sigma^2)', \nu')$ of X with respect to Q are*

$$\gamma' = \gamma + \beta\sigma + \int h(x) \; (y(x) - 1) \; \nu(dx),$$

$$(\sigma^2)' = \sigma^2,$$

$$\nu'(dx) = y(x) \nu(dx).$$

Moreover, if $Q \sim P$ is structure preserving with Girsanov parameters (β, y), then the density process of Q with respect to P is given by $Z = \mathcal{E}(M)$ with

$$M = \beta X^c + (y - 1) * \left(\mu^X - \nu^X \right). \tag{4.8}$$

If, on the other hand, we define a probability measure Q by $dQ/dP = \mathcal{E}(M)_T$ where M satisfies (4.8) with Girsanov parameters (β, y), then Q is structure preserving.

For the proof of these results we refer to Raible (2000), Proposition 2.19, and Hubalek & Sgarra (2006), Theorem 5.

4.2 Stochastic volatility models

In the Samuelson model, stock prices are modelled under the risk-neutral measure as geometric Brownian motion with constant volatility $\sigma > 0$. A more complex class of models can then be generated by modelling σ as a stochastic process rather than a constant. This and the following sections serve as an introduction to these 'stochastic volatility' (SV) models. Our selection of topics is motivated by the fact that SV models are much more complex than exponential Lévy models, and give rise to a whole range of new phenomena. With SV models, the distinction between local martingales and martingales is a recurrent topic. The price process itself can be a strict local martingale under the risk-neutral measure which is related to the occurrence of financial bubbles. To show that stochastic exponentials in an SV framework are martingales can be rather demanding as more standard criteria such as Novikov's criterion often do not work. Moreover, the filtration generated by the price process and the one generated by the driving noise processes need not coincide.

In general, SV models generate incomplete markets. However, in the presence of a liquid market in plain vanilla options, and in a diffusion process framework, one sometimes can complete the market by trading in both the stock and an option written on the stock. Assuming constant coefficients, this leads to a PDE in two state variables generalising the Black-Scholes PDE from which one can derive the replicating hedging strategy. This approach works out when option prices are convex in the asset price level, and this property can be studied via coupling methods.

4.2.1   *Examples*

(A) Heston model

The stochastic logarithm of the price process X and the volatility process V are given as weak solutions (see the following section) to the following system of stochastic differential equations:

$$dX_t = \left(\mu + \delta V_t^2\right) dt + V_t \, dB_t,$$
$$dV_t = (\kappa - \lambda V_t) \, dt + \sigma \sqrt{V_t} \, dW_t.$$

Here $\mu, \delta, \kappa \geq 0, \lambda > 0, \sigma > 0$ are constants, $V_0 > 0$, and B, W are Brownian motions with constant correlation $\rho \in [-1, 1]$. Due to the square-root term in front of W, the volatility process V is non-negative. Observing X, we can also observe its quadratic variation $[X]$, and therefore the volatility process V as well since $V^2 = d[X]/dt$. Noting that $\mathbb{F}^V = \mathbb{F}^W$, it follows that $\mathbb{F}^X = \mathbb{F}^{B,W}$.

(B) Stein & Stein model

With the same specifications as under (A), consider

$$dX_t = \left(\mu + \delta V_t^2\right) dt + V_t \, dB_t,$$
$$dV_t = (\kappa - \lambda V_t) \, dt + \sigma \, dW_t.$$

V is an Ornstein-Uhlenbeck process, so can take negative values. By $V^2 = d[X]/dt$, one can observe the modulus of V. However, it can be shown that we cannot observe $sign(V)$, so $\mathbb{F}^X \subset \mathbb{F}^{B,W}$, with a strict inclusion. Moreover, it has been shown in Rheinländer (2005) that $\mathbb{F}^X$ is not even immersed in $\mathbb{F}^{B,W}$. Recall that a filtration $\mathbb{F} \subset \mathbb{G}$ is **immersed** into $\mathbb{G}$ when all $\mathbb{F}$-martingales remain martingales in the larger filtration $\mathbb{G}$. For the purpose of option pricing and hedging one cannot appeal to martingale representations with respect to the Brownian filtration $\mathbb{F}^{B,W}$, and instead one must work within the complicated filtration $\mathbb{F}^X$.

(C) Barndorff-Nielsen & Shephard (BNS) model

This model has the structure

$$dX_t = (\mu + \delta V_t) \, dt + \sqrt{V_t} \, dB_t + \rho \, dZ_t,$$
$$dV_t = -\lambda V_t \, dt + dZ_t.$$

Here $\mu, \delta, \lambda > 0, \rho$ are constants, and we assume that $V_0 > 0$. B denotes a Brownian motion, and Z a subordinator. In the BNS model, the volatility process increases by jumps only, and in between jumps decreases exponentially with rate λ. Often ρ is chosen to be negative, to incorporate a 'leverage effect' whereby an increase in volatility is associated with downwards stock movements. The subordinator Z can be observed via the jump process ΔX, and the volatility process V via the quadratic variation $[X]$. Since $\mathbb{F}^V = \mathbb{F}^Z$, it follows that $\mathbb{F}^X = \mathbb{F}^{B,Z}$.

4.2.2 *Stochastic differential equations and time change*

Consider two measurable functions α, β with values in $d \times d$ matrices respectively d-vectors, and the autonomous stochastic differential equation (SDE)

$$dX_t = \alpha(X_t)\, dB_t + \beta(X_t)\, dt. \tag{4.9}$$

A **weak solution** is a pair (X, B) of d-dimensional adapted processes defined on a filtered probability space $(\Omega, \mathcal{G}, \mathbb{G}, Q)$ such that B is a standard d-dimensional Q-Brownian motion, and X satisfies for $i = 1, \ldots, d$,

$$X_t^{(i)} = X_0^{(i)} + \sum_j \int_0^t \alpha_{ij}(X_s)\, dB_s^{(j)} + \int_0^t \beta_i(X_s)\, ds.$$

There is **uniqueness in law** if, whenever (X, B) and (X', B') are two weak solutions with the same initial condition, the laws of X and X' are equal. There is **pathwise uniqueness** if (X, B) and (X', B') are defined on the same filtered space with $B = B'$ and X and X' are indistinguishable. A weak solution (X, B) is said to be a **strong solution** if X is adapted to the completed filtration $\mathbb{F}^B$ generated by B. Pathwise uniqueness implies uniqueness in law and that every solution is strong, see Theorem IX.1.7 of Revuz & Yor (1999).

We can construct weak solutions via the technique of time change. For discussing this, consider a filtered probability space $(\Omega, \mathcal{F}, \mathbb{F}, P)$ and the time interval $[0, \infty)$.

Theorem 4.22. *Dambis-Dubins-Schwarz (DDS)*. *Let M be a continuous $(P, \mathbb{F})$-local martingale vanishing at zero and such that $[M]_\infty = \infty$, and set*

$$A_t = \inf\{s \,|\, [M]_s > t\}. \tag{4.10}$$

Then B defined by $B_t = M_{A_t}$ is an $\mathbb{F}^A$-Brownian motion, where $\mathcal{F}_t^A = \mathcal{F}_{A_t}$, and $M_t = B_{[M]_t}$.

Proof.   See Revuz & Yor (1999), Theorem V.1.6.   □

Definition 4.23. Let M be a continuous $(P, \mathbb{F})$-martingale vanishing at zero and such that $[M]_\infty = \infty$, and consider its DDS representation $M = B_{[M]}$. The process M is called an **Ocone martingale** if B and $[M]$ are independent; it is called a **pure martingale** if $[M]_t$ is $\mathcal{F}^B_\infty$-measurable for every $t \geq 0$.

Proposition 4.24. *Pure martingales generate complete markets.* *Let M be a pure martingale. Then the set of equivalent martingale measures for M on $\mathcal{F}^M_\infty$ consists of a singleton.*

Proof.   Let $P \sim Q$ be two martingale measures for M. As the quadratic variation $[M]$ is invariant under equivalent measure changes, the DDS Brownian motion of M, denoted B, is invariant as well. Therefore P and Q agree on $\mathcal{F}^B_\infty$ which contains $\mathcal{F}^M_\infty$ by the purity of M.   □

Examples for pure martingales are the stochastic integrals $\int B^n \, dB$ for $n \in \mathbb{N}$ and a Brownian motion B, which has been shown by Beghdadi-Sakrani (2003). The following result is Theorem 2 in Vostrikova & Yor (2000).

Proposition 4.25. *Ocone martingales i.g. generate incomplete markets.* *Let M be an Ocone martingale. Then the space of square-integrable $\mathbb{F}^M$-martingales is the direct sum of the stable space generated by the $\mathbb{F}^{[M]}$-martingales and the stable space generated by M.*

An example of an Ocone martingale is $\int W \, dB$ where B and W are independent Brownian motions. Ocone martingales are never pure unless they are Gaussian, in which case the generated market is complete.

4.2.3   *Construction of a solution via coupling*

Let B and W be two, possibly correlated, Brownian motions. Initially we consider a time interval $[0, \infty)$ which facilitates time change methods but afterwards restrict our attention to a compact interval $[0, T]$ in line with the rest of the book. Let $\sigma > 0, \alpha, \beta$ be measurable functions. We moreover assume that σ is continuous. Consider the following SDE for the pair of price/volatility processes:

$$dS_t = \sigma(S_t) V_t \, dB_t, \qquad\qquad (4.11)$$
$$dV_t = \alpha(V_t) \, dW_t + \beta(V_t) \, dt$$

where $S_0 = s > 0$, $V_0 = v \geq 0$, and

$$d\,[B,W]_t = \rho\,(V_t)\,dt$$

for a measurable function ρ taking values in $(-1,1)$. Moreover, we consider the auxiliary SDE

$$dX_t = \sigma\,(X_t)\,dB_t^X, \qquad\qquad (4.12)$$

$$dY_t = \frac{\alpha\,(Y_t)}{Y_t}\,dW_t^Y + \frac{\beta\,(Y_t)}{Y_t^2}\,dt$$

where $X_0 = s$, $Y_0 = v$, and B^X, W^Y are two Brownian motions with correlation

$$d\left[B^X, W^Y\right]_t = \rho\,(Y_t)\,dt.$$

We let $\mathbb{G}$ be the augmented filtration generated by B^X and W^Y.

Theorem 4.26. *Existence of weak solution via coupling.* *Assume that the SDE (4.12) has a unique strong solution (X,Y) which does not explode. Let $\Gamma = \int Y^{-2}\,dt$ and $A = \Gamma^{-1}$, and assume that Γ also does not explode. Define the processes B, W via*

$$B_t = \int_0^{A_t} \frac{dB_s^X}{Y_s}, \qquad W_t = \int_0^{A_t} \frac{dW_s^Y}{Y_s}.$$

Then B, W are $\mathbb{F}$-Brownian motions where $\mathcal{F}_t := \mathcal{G}_{A_t}$. Moreover, set $S_t = X_{A_t}$ and $V_t = Y_{A_t}$. Then the pair $((S,V),(B,W))$ is a weak solution to the SDE (4.11).

Proof. The time change A is well-defined since Γ is strictly increasing, continuous and non-exploding. Set

$$M_t = \int_0^t \frac{dW_s^Y}{Y_s}$$

and

$$W_t = M_{A_t} = \int_0^{A_t} \frac{dW_s^Y}{Y_s} = \int_0^t \frac{dW_{A_s}^Y}{Y_{A_s}}.$$

Then, as $[M] = \Gamma$, it follows by the Dambis-Dubins-Schwarz theorem that W is an $\mathbb{F}$-Brownian motion. Moreover, (V, W) solves

$$dV_t = \alpha\,(Y_{A_t})\frac{dW_{A_t}^Y}{Y_{A_t}} + \beta\,(Y_{A_t})\frac{dA_t}{Y_{A_t}^2}$$

$$= \alpha\,(V_t)\,dW_t + \beta\,(V_t)\,dt.$$

Similarly, B is an $\mathbb{F}$-Brownian motion and (S, B) solves

$$dS_t = dX_{A_t} = \sigma(X_{A_t})\, dB_{A_t}^X = \sigma(S_t) V_t\, dB_t.$$

Finally,

$$[B, W]_t = \int_0^{A_t} \frac{d[B^X, W^Y]_s}{Y_s^2} = \int_0^{A_t} \rho(Y_s)\, d\Gamma_s = \int_0^t \rho(V_s)\, ds. \qquad \Box$$

4.2.4 Convexity of option prices

Next we study whether option prices are convex in the asset price level. Consider an SV model as in (4.11), together with an auxiliary model as in (4.12). Given a convex payoff function h, which is not affine linear, we define the corresponding European option price as

$$\phi(s, v) = E\left[h(S_T) | S_0 = s, V_0 = v\right].$$

Theorem 4.27. Convexity of option prices. *With the assumptions of Theorem 4.26 in place, suppose that $\phi(s, v)$ is finite for all positive initial starting points. Moreover, assume that either $\rho = 0$ and that S is a martingale, or that $\sigma(s) = \sigma s$ for a constant $\sigma > 0$. Then for each fixed v, $\phi(s, v)$ is strictly convex in s.*

Proof. Firstly, if $\rho = 0$ then S can be written as $S_t = X_{A_t}$ where X and A are independent. By conditioning on A, and noting that the convexity property is maintained when averaging over A_T, we can reduce the model to

$$dX_t = \sigma(X_t)\, dB_t.$$

Let $0 < z < y < x$, and for three independent Brownian motions α, β, γ define X, Y, Z as weak solutions to

$$dX_t = \sigma(X_t)\, d\alpha_t, \quad X_0 = x,$$
$$dY_t = \sigma(Y_t)\, d\beta_t, \quad Y_0 = y,$$
$$dZ_t = \sigma(Z_t)\, d\gamma_t, \quad Z_0 = z.$$

The first intersection times are defined as $H_x = \{t \geq 0\, |X_t = Y_t\}$ and $H_z = \{t \geq 0\, |Y_t = Z_t\}$, and we set $\tau = H_x \wedge H_z \wedge T$. We have by symmetry

$$E\left[(X_T - Z_T) h(Y_T) \mathbb{I}_{\tau = H_x}\right] = E\left[(Y_T - Z_T) h(X_T) \mathbb{I}_{\tau = H_x}\right],$$

and moreover

$$E\left[(X_T - Y_T) h(Z_T) \mathbb{I}_{\tau = H_x}\right] = 0.$$

An analogous relation holds on $\{\tau = H_z\}$. Furthermore, on $\{\tau = T\}$ we have $Z_T < Y_T < X_T$, and by convexity of h

$$(X_T - Z_T) h(Y_T) \leq (X_T - Y_T) h(Z_T) + (Y_T - Z_T) h(X_T).$$

Using $\Omega = \{\tau = H_x\} \cup \{\tau = H_z\} \cup \{\tau = T\}$ and taking expectations yields

$$E[(X_T - Z_T) h(Y_T)] \leq E[(X_T - Y_T) h(Z_T)] + E[(Y_T - Z_T) h(X_T)].$$

It follows by the independence of X, Y and Z that for $\lambda = (x - y)/(x - z)$

$$E[h(Y_T)| \mathcal{F}_t] \leq \lambda E[h(Z_T)| \mathcal{F}_t] + (1 - \lambda) E[h(X_T)| \mathcal{F}_t].$$

Secondly, if $\sigma(s) = \sigma s$ but not necessarily $\rho = 0$ then we can write $S = sZ$ where

$$Z_t = \exp\left(\sigma B^X_{A_t} - \frac{1}{2}\sigma^2 A_t\right),$$

so that Z does not depend on s. It follows that for any convex combination $\sigma s = \lambda q + (1 - \lambda) r$, $\lambda \in (0, 1)$, we have

$$h(\sigma s Z_T) \leq \lambda h(q Z_T) + (1 - \lambda) h(r Z_T)$$

by the convexity of h. Taking expectations yields the result since the inequality is strict on a non-zero set since h is not affine-linear. $\qquad \square$

Remark. There is a counter-example in Hobson (2010) which shows that even when S is a true martingale, the convexity property might fail.

4.2.5 *Market completion by trading in options*

In this subsection we denote the volatility process by Y since we reserve V for the value process associated with an option. We let $\sigma > 0$ be constant, $\alpha > 0, \beta, \rho$ be measurable functions, and consider the SDE

$$dS_t = \sigma S_t Y_t \, dB_t, \qquad\qquad (4.13)$$
$$dY_t = \alpha(Y_t) \, dW_t + \beta(Y_t) \, dt$$

where

$$d[B, W]_t = \rho(Y_t) \, dt, \quad \rho \in (-1, 1).$$

We assume that a weak solution exists, that Y is positive, and non-zero a.e. For the purpose of martingale representation we introduce a Brownian motion Z orthogonal to B such that

$$W_t = \rho(Y_t) B_t + \sqrt{1 - \rho^2(Y_t)} Z_t,$$

and re-write the SDE (4.13) as

$$dS_t = \sigma S_t Y_t \, dB_t,$$
$$dY_t = \vartheta\,(Y_t)\, dB_t + \eta\,(Y_t)\, dZ_t + \beta\,(Y_t)\, dt,$$

with $\vartheta = \alpha\rho$, $\eta = \alpha\sqrt{1 - \rho^2}$. The completed filtration generated by (B, Z) is denoted $\mathbb{F} = (\mathcal{F}_t)$, and we assume that $\mathbb{F} = \mathbb{F}^S$, so that the value process of an option can be obtained by conditioning with respect to $\mathbb{F}$.

If we are only allowed to trade in S, then in general, the resulting market is incomplete due to the presence of two uncorrelated sources of noise but only one tradeable asset. When there is a liquidly traded vanilla option written on S available, with value process V, it is sometimes possible to complete the market by trading in both S and V. An extension of the notion of completeness defined in Section 3.2 should be used, since our price process (S, V) is vector-valued. A claim is attainable if it can be written as sum of a constant and stochastic integrals of admissible strategies with respect to S and V. A market is called complete if all bounded claims are attainable in this generalised sense.

Consider a put option with strike K and maturity T, then V is given as the martingale

$$V_t = E\left[\left(K - S_T\right)^+ \Big| \mathcal{F}_t\right].$$

Since the put has a bounded payoff, we have $V \in \mathcal{M}^2$. By Brownian martingale representation, there exist $\vartheta^V \in L^2(B)$, $\eta^V \in L^2(Z)$ such that

$$dV = \vartheta^V \, dB + \eta^V \, dZ.$$

We assume that there exists a smooth function $v = v(t, s, y)$ such that

$$V_t = v(t, S_t, Y_t).$$

By Itô's formula, denoting partial derivatives as subscripts,

$$dv = \left(v_t + \beta v_y + \frac{1}{2}\sigma^2 S^2 Y^2 v_{ss} + \sigma SY \vartheta v_{ys} + \frac{1}{2}\left(\vartheta^2 + \eta^2\right) v_{yy}\right) dt$$
$$+ \left(\sigma SY v_s + \vartheta v_y\right) dB + \eta v_y \, dZ.$$

Hence we can identify

$$\vartheta^V = \sigma SY v_s + \vartheta v_y, \quad \eta^V = \eta v_y. \tag{4.14}$$

The drift term must vanish, so we derive the PDE

$$v_t + \beta v_y + \frac{1}{2}\sigma^2 s^2 y^2 v_{ss} + \sigma sy\vartheta v_{ys} + \frac{1}{2}\left(\vartheta^2 + \eta^2\right) v_{yy} = 0, \tag{4.15}$$
$$v\,(T, s, y) = (K - s)^+.$$

Moreover, consider an $\mathcal{F}_{T'}$-measurable, square-integrable claim H maturing at $T' < T$, with associated value process V^H having the martingale representation

$$dV^H = \vartheta^H\, dB + \eta^H\, dZ,$$

for $\vartheta^H \in L^2(B)$, $\eta^H \in L^2(Z)$.

If the market were complete, we could find admissible strategies ϕ, ψ such that $\int \phi\, dS$ and $\int \psi\, dV$ are square-integrable martingales and

$$dV^H = \phi\, dS + \psi\, dV.$$

In the case when v_y, and then, due to (4.14) together with $\eta \neq 0$, η^V never vanish, we can solve for ϕ, ψ as

$$\phi = \frac{\vartheta^H - \psi\vartheta^V}{\sigma SY}, \quad \psi = \frac{\eta^H}{\eta^V}.$$

To sum up, the market given by trading in (S, V) is complete if

$$v_y = \frac{\partial v}{\partial y} \neq 0. \tag{4.16}$$

To find out when this condition holds, we assume that we can differentiate the PDE (4.15) satisfied by the option with respect to y. This yields the following PDE for $u = \partial v/\partial y$:

$$u_t + \beta_y u + (y\vartheta_y + \vartheta)\,\sigma s u_s + (\beta + \vartheta\vartheta_y + \eta\eta_y)\,u_y \tag{4.17}$$

$$+\frac{1}{2}\sigma^2 s^2 y^2 u_{ss} + \sigma s y \vartheta u_{sy} + \frac{1}{2}\left(\vartheta^2 + \eta^2\right) u_{yy} = \sigma^2 s^2 y v_{ss},$$

$$u\,(T, s, y) = 0.$$

By Theorem 4.27, the option price function v is strictly convex in s, i.e. $v_{ss} > 0$. If the coefficients are sufficiently regular, we can apply the strong maximum principle as in Friedman (1975) to get that $u\,(t, s, y)$ has the same sign as y. By assumption, $Y > 0$ a.e., and therefore the completeness condition (4.16) holds.

4.2.6   *Bubbles and strict local martingales*

Stochastic volatility models lead to incomplete markets that exhibit new phenomena which are absent in the Lévy world. In particular, bubbles are a way in which local martingales can impact the behaviour of financial markets. Bubbles in the sense discussed in this section are consistent with the first fundamental theorem of asset pricing, so not allow for arbitrage, but still generate many unusual and unexpected phenomena. In this section the dynamics S of the discounted stock price process are defined directly under a given and fixed martingale measure Q.

Definition 4.28. The price process S is said to exhibit a **bubble** if under the equivalent martingale measure Q, S is a local martingale but not a martingale. We refer to such processes as **strict local martingales**.

In case of price processes which are strict local martingales one has to take care when choosing a set of admissible strategies when hedging. For example, if the set of admissible integrands is

$$\Theta = \left\{ \vartheta \in L\left(S\right) \ \middle| \ \int \vartheta \, dS \text{ is a } Q\text{-martingale for all } Q \in \mathcal{M}^e \right\}$$

and the price process S is a positive strict local martingale under some $Q \in \mathcal{M}^e$, hence a Q-super-martingale by Proposition 2.6, then the strategy 'buy and hold' one stock, namely $\vartheta \equiv 1$, is not admissible! Although this might seem odd at first glance, it makes sense if we consider that when the stock price will certainly fall on average, it is indeed not sensible to hold the stock.

Proposition 4.29. *Characterisation of bubbles.* S *exhibits a bubble if and only if*

$$\limsup_n n \, Q \left(\sup_{0 \leq t \leq T} S_t > n \right) > 0.$$

Proof. See Azéma *et al.* (1980). □

This characterisation illustrates that the definition of a bubble coincides with our intuition that during a bubble phase very large stock prices will be attained with significant probability, since the probability that the stock price will surpass the level n is of the order $1/n$.

Bubbles defined in this way are also a mathematical explanation for scenarios where the commonly understood 'rules of thumb' of derivatives pricing are no longer true. Let us first present a result about local martingales, see Cox & Hobson (2005), Theorem A.1.

Lemma 4.30. *Assume that S is a continuous non-negative Q-local martingale, with $S_0 = 1$. Let $h : \mathbb{R}^+ \to \mathbb{R}$ be a convex function such that*

$$\limsup_{x \to \infty} h(x)/x = c \in [0, \infty].$$

Then

$$\sup_{\tau} E_Q\left[h\left(S_\tau\right)\right] = E_Q\left[h\left(S_T\right)\right] + c\left(T - E_Q\left[S_T\right]\right)$$

where the supremum is taken over all stopping times $0 \le \tau \le T$.

The next theorem illustrates many of the quirky features exhibited by markets with bubbles.

Theorem 4.31. Bubble phenomena. *The local martingale S has a bubble if and only if any of the following conditions hold:*

(i) *S is a strict local martingale;*
(ii) *The forward price is below the current price, i.e.*

$$E_Q\left[S_T \mid \mathcal{F}_t\right] < S_t;$$

(iii) *Put-call parity breaks down, i.e.*

$$E_Q\left[(S_T - K)^+ \mid \mathcal{F}_t\right] - E_Q\left[(K - S_T)^+ \mid \mathcal{F}_t\right] < S_t - K;$$

(iv) *For some strike prices, $K > 0$, the price of a European call option is lower than the price of an American call with the same strike, i.e.*

$$E_Q\left[(S_T - K)^+ \mid \mathcal{F}_t\right] < \operatorname*{ess\,sup}_{\tau \le T} E_Q\left[(S_\tau - K)^+ \mid \mathcal{F}_t\right];$$

(v) *The value of a European put option exceeds the value of a corresponding future, i.e.*

$$\lim_{K \uparrow \infty} E_Q\left[(K - S_T)^+ \mid \mathcal{F}_t\right] - K + S_t > 0;$$

(vi) *An American call option retains some of its value as the strike is increased, i.e.*

$$\lim_{K \uparrow \infty} \left(\operatorname*{ess\,sup}_{\tau \le T} E_Q\left[(S_\tau - K)^+ \mid \mathcal{F}_t\right]\right) > 0.$$

Proof. The first condition holds by definition. As the stock price is assumed to be bounded from below by zero, Fatou's lemma implies that S is a strict super-martingale which is condition (ii). Condition (iii) follows from applying (ii) when taking the conditional Q-expectation of

$$(S_T - K)^+ - (K - S_T)^+ = S_T - K. \tag{4.18}$$

Also, (v) is equivalent to (ii) once we observe that

$$E_Q\left[(K - S_T)^+ \mid \mathcal{F}_t\right] - K + S_t = E_Q\left[(S_T - K)^+ - S_T \mid \mathcal{F}_t\right] + S_t,$$

and that the European call option becomes worthless as the strike is increased, i.e.

$$\lim_{K \uparrow \infty} E_Q \left[(S_T - K)^+ | \mathcal{F}_t \right] = 0. \qquad (4.19)$$

Next, consider the convex function $h(x) = (x - K)^+$ which has the property that $\limsup_{x \to \infty} h(x)/x = 1$. By applying Lemma 4.30, we observe that

$$\operatorname*{ess\,sup}_{\tau \leq T} E_Q \left[(S_\tau - K)^+ | \mathcal{F}_t \right] - E_Q \left[(S_T - K)^+ | \mathcal{F}_t \right] = S_t - E_Q \left[S_T | \mathcal{F}_t \right],$$

which proves that condition (ii) is equivalent to condition (iv) holding for an arbitrary $K > 0$. Therefore, (4.19) and condition (iv) are equivalent to (vi). $\qquad \square$

We conclude this section with some examples of stock price processes containing bubbles, and are prone to the behaviour described in the previous theorem.

(A) Inverse Bessel process

The prototypical example of a strict local martingale on an infinite time horizon $(T = \infty)$ is the inverse Bessel process. Let $B = (B^1, B^2, B^3)$ be a three-dimensional Brownian motion with initial point $B_0 = (1, 0, 0)$ and consider the function $u : \mathbb{R}^3 \backslash \{(0,0,0)\} \to \mathbb{R}^+$ defined by $u(x) = 1/\|x\|$, where $\|\cdot\|$ is the Euclidean norm. The process $S = u(B)$ is a positive local martingale but not a true martingale because $\lim_{t \to \infty} E[S_t] = 0$. In fact, if S is a strong solution to

$$dS_t = f(S_t) \, dB_t$$

where $S_0 = 1$ and the function f satisfies $\int_x^\infty 1/f^2(y) \, dy < \infty$, for some $x \in (0, 1)$, then the process S is a strict local martingale on the interval $[0, \infty]$. The inverse Bessel process is an example of a process of this type with $f(y) = -y^2$. This example shows also that bubbles can arise even in a complete market.

(B) SV models, incomplete market case

Consider the stock price S defined on the finite horizon $[0, T]$ to be the solution of the following SDE, under a martingale measure Q:

$$dS_t = S_t \sigma_t \, dB_t,$$

where for some constants α, β, the volatility σ solves

$$d\sigma_t = \sigma_t \left(\alpha \, dt + \beta \, dW_t\right).$$

Both B and W are correlated Brownian motions with $d[B, W] = \rho \, dt$ where $|\rho| \in (0, 1)$. Notice that the Brownian motion driving the volatility can be decomposed as $W = \rho B + \rho^\perp B^\perp$ where $\rho^\perp = \sqrt{1 - \rho^2}$ and the Brownian motion $B^\perp$ is orthogonal to B. The SDE for the volatility is now

$$d\sigma_t = \alpha\sigma_t \, dt + \beta\sigma_t \left(\rho \, dB_t + \rho^\perp \, dB_t^\perp\right).$$

We will prove that this price process exhibits a bubble by arguing by contradiction. Assume that S is a true Q-martingale and define a new probability measure $\widetilde{Q}$ by setting $d\widetilde{Q}/dQ = S_T/S_0$. Then Girsanov's theorem implies that the processes $\widetilde{B} = B - \int \sigma \, dt$ and $B^\perp$ are both $\widetilde{Q}$-Brownian motions. Under this new measure $\widetilde{Q}$ the volatility solves the following SDE:

$$d\sigma_t = \left(\alpha\sigma_t + \beta\rho\sigma_t^2\right) dt + \beta\sigma_t \left(\rho \, d\widetilde{B} + \rho^\perp \, dB^\perp\right).$$

When $\beta\rho > 0$ this SDE will explode to infinity within finite time. As the volatility process does not explode under the original measure Q we can conclude that $\widetilde{Q} \nsim Q$ which implies S/S_0 cannot be a martingale. Hence, if $\beta\rho > 0$ the stock price S is a strict local martingale.

4.2.7   *Stochastic exponentials*

In the framework of Brownian SV models, often Novikov's criterion is not sufficient to verify whether certain stochastic exponentials are true martingales. In such cases the following approach is quite useful. See Liptser & Shiryaev (2000), Section 6.2, Example 3.

Proposition 4.32. *Liptser and Shiryaev's criterion.* *Let T be a finite time horizon, V a predictable process with*

$$P\left(\int_0^T V_s^2 \, ds < \infty\right) = 1,$$

and B a Brownian motion. Provided that there is $\varepsilon > 0$ such that

$$\sup_{0 \leq t \leq T} E\left[\exp\left(\varepsilon V_t^2\right)\right] < \infty,$$

then the stochastic exponential $\mathcal{E}\left(\int V \, dB\right)$ is a martingale on $[0, T]$.

4.3 Notes and further reading

A succinct introduction to Lévy processes can be found in Protter (2005). Applebaum (2009) and Kyprianou (2006) are introductory textbooks about Lévy processes, whereas Sato (1999) offers a comprehensive analytical treatment without reference to martingales. Time-changed Lévy processes, which we cover only marginally, are one subject of Barndorff-Nielsen & Shiryaev (2011). Lévy processes in finance are treated in Cont & Tankov (2003), as well as in Schoutens (2003). Generalised hyperbolic distributions are covered in much detail by Raible (2000), whereas for the time-inhomogenous case we refer to Kluge (2005).

A very good survey about stochastic volatility methods and semi-martingale characteristics complementing this chapter is Kallsen (2006). Affine stochastic volatility models have been studied by Duffie *et al.* (2003). The coupling method in more generality can be found in Hobson (2010), convexity of option prices is explored in Hobson (1998), and the section about market completion is adapted from Renault & Touzi (1997). Strict local martingales have been linked to bubbles by Cox & Hobson (2005). See Jarrow *et al.* (2010) for a recent treatment.

Non-Gaussian Ornstein-Uhlenbeck processes are the basic modelling tool for energy and electricity markets in the monograph by Benth *et al.* (2008).

Chapter 5

Static Hedging

Static hedging is the replication of a derivative with a complicated payoff using a portfolio of simpler ones. For instance, European options may be replicated by plain vanilla put and call options, barrier options by European options, or an Asian option with a floating strike can be replicated by one with a fixed strike. This approach typically assumes that there is no trading allowed after inception. In the case of path-dependent options, a semi-static hedging may be appropriate, whereby trading is allowed at discrete time points such as at the hitting time of barriers.

The modern alchemy of static hedging relies crucially on both the concept of the dual process and on the symmetry properties of the underlying. These concepts are discussed in the framework of exponential Lévy processes and continuous stochastic volatility models. In the case of non-trivial carrying costs like interest and dividend rates, one has to consider quasi self-dual processes. These are of particular interest when it comes to the semi-static hedging of barrier options.

5.1 Static hedging of European claims

A European **call option** on an asset S with maturity T and strike price K gives its holder the right, but not the obligation to buy the asset at time T for the fixed price K. The payoff of a call is therefore

$$H = (S_T - K)^+ .$$

In contrast, the payoff of a **put option** is given as $(K - S_T)^+$, and that of a **forward contract** equals $S_T - K$. In the following we work with discounted quantities. It turns out that we can express the payoff of a call in terms of a forward and a put. This is due to the equality

$$(S_T - K)^+ = (S_T - K) + (K - S_T)^+. \tag{5.1}$$

If we can find a probability measure Q such that the price process S is a Q-martingale, then we get

$$E_Q \left[(S_T - K)^+ \Big| \mathcal{F}_t \right] = (S_t - K) + E_Q \left[(K - S_T)^+ \Big| \mathcal{F}_t \right]. \tag{5.2}$$

This relation is called **put-call parity**, and illustrates that the value of a call is at all times the sum of the value of a forward and a put. When S is a Q-martingale, this relation allows us to infer call prices from put prices and vice verse. However, as we have seen in Theorem 4.31, put-call parity in the form of (5.2) fails in the case when S is a strict local martingale. Put-call parity always holds if we replace the European call by a corresponding American call.

Proposition 5.1. *Put-call parity.* *Let S be a local Q-martingale. Then*

$$\sup_{\tau \leq T} E_Q \left[(S_\tau - K)^+ | \mathcal{F}_t \right] = (S_t - K) + E_Q \left[(K - S_T)^+ \Big| \mathcal{F}_t \right]. \tag{5.3}$$

Proof. It follows from Lemma 4.30 that the difference between American and European call price is

$$\sup_{\tau \leq T} E_Q \left[(S_\tau - K)^+ | \mathcal{F}_t \right] - E_Q \left[(S_T - K)^+ | \mathcal{F}_t \right] = S_t - E_Q \left[S_T | \mathcal{F}_t \right],$$

which is equal to zero if S is a Q-martingale. Furthermore, by the relation (5.1) between the payoffs of puts and calls we have that

$$E_Q \left[(S_T - K)^+ | \mathcal{F}_t \right] = E_Q \left[S_T | \mathcal{F}_t \right] - K + E_Q \left[(K - S_T)^+ \Big| \mathcal{F}_t \right].$$

Combining these two equations yields the put-call parity (5.3). $\square$

Calls, puts and forwards can be considered in a way to be the atoms of more general European claims. For instance, many popular European options like straddles or butterfly spreads can be created as a linear combination of these. Consider for example a straddle which is a combination of a call and a put with the same strike price and maturity. The straddle gains profit the more S_T is away from the strike price, symmetrically in both directions. On the other hand, for S_T being close to the strike, the straddle generates a net loss (since one has to pay a price for entering it). Therefore, the straddle can be seen as a bet on high future volatility.

A general European option pays its holder an amount

$$H = h(S_T)$$

at maturity, where we assume that h is a sufficiently smooth function. It is always possible to describe a replicating portfolio for the option in terms of forwards, puts and calls. In general, this requires purchasing a portfolio of vanillas with a continuum of strikes, but this can be readily approximated using options with a finite number of strikes.

Proposition 5.2. *Taylor representation of European options.* *For any $x \geq 0$, any twice continuously differentiable European payoff h can be written as*

$$h(S_T) = h(x) + h'(x)(S_T - x) \tag{5.4}$$
$$+ \int_0^x h''(K)(K - S_T)^+ \, dK + \int_x^\infty h''(K)(S_T - K)^+ \, dK.$$

Proof. Apply Taylor's formula with the remainder term in integral form,

$$h(S_T) = h(x) + h'(x)(S_T - x) + \int_x^{S_T} h''(K)(S_T - K) \, dK,$$

and now use $S_T - K = (S_T - K)^+ - (K - S_T)^+$. □

5.2 Duality principle in option pricing

5.2.1 *Dynamics of the dual process*

Definition 5.3. Let $S = \exp(X)$ be a martingale with $E[S_T] = 1$. We define the probability measure P' by

$$\frac{dP'}{dP} = S_T.$$

The **dual process** S' is

$$S' = \frac{1}{S} = \exp(-X).$$

By Bayes' formula, S' is a martingale with respect to the **dual measure** P'.

Proposition 5.4. *Dual Lévy triplet.* *Let $S = \exp(X)$ for a Lévy process X, $X_0 = 0$, which satisfies (EM) and has triplet (γ, σ^2, ν). Assume that S is a martingale. Then the triplet of $-X$ under the dual measure P' is*

$$\gamma' = -\gamma - \sigma - \int x(e^x - 1) \, \nu(dx),$$
$$\left(\sigma^2\right)' = \sigma^2,$$
$$\nu'(dx) = e^{-x} \nu(-dx).$$

Proof.   Since $\exp(X)$ is a positive P-martingale, there exists Y such that $\exp(X) = \mathcal{E}(Y)$. By the martingale condition (4.6), Y is a Lévy process with decomposition

$$Y = X^c + (e^x - 1) * \left(\mu^X - \nu^X\right).$$

By Theorem 4.21 about structure preserving measure transforms, the measure change from P to P' can therefore be described by the Girsanov parameters $\beta = 1$, $y(x) = \exp(x)$, and the triplet of X with respect to P' is

$$\gamma^* = \gamma + \sigma + \int x\,(e^x - 1)\,\nu\,(dx),$$
$$\left(\sigma^2\right)^* = \sigma^2,$$
$$\nu^*\,(dx) = e^x\,\nu\,(dx).$$

The expressions for γ' and $\left(\sigma^2\right)'$ follow immediately, and the Lévy measure ν' of $-X$ under P' can be determined for every Borel set Γ by

$$\mathbb{I}_\Gamma(x)\,\nu'(dx) = \mathbb{I}_\Gamma(-x)\,\nu^*(dx)$$
$$= \mathbb{I}_\Gamma(-x)\,e^x\,\nu(dx)$$
$$= \mathbb{I}_\Gamma(x)\,e^{-x}\,\nu(-dx).$$

$\square$

Proposition 5.5. *Dual decomposition of continuous martingales.*
Let $S = \exp(X)$ be a continuous martingale with $X_0 = 0$. Then the canonical decomposition of $-X$ under the dual measure P' is

$$-X = \left(-X + \frac{1}{2}\,[X]\right) - \frac{1}{2}\,[X]$$

where $-X + \frac{1}{2}\,[X]$ is a local P'-martingale.

Proof.   Since S is a positive continuous martingale, we can write $S = \mathcal{E}(Y)$ with $Y = X + \frac{1}{2}\,[X]$. In particular, $[Y] = [X]$, so by Girsanov's theorem,

$$Y - \int \frac{1}{S}\,d\,[S,Y] = Y - [Y] = X - \frac{1}{2}\,[X]$$

is a local P'-martingale from which the statement follows. $\square$

5.2.2 *Duality relations*

The standing assumption of this section is

$$\boxed{S = \exp(X) \text{ is a martingale with } S_0 = 1.}$$

In the following we denote by

$$\mathbb{C}_T\left(S; K\right) = E\left[\left(S_T - K\right)^+\right],$$
$$\mathbb{P}_T\left(S; K\right) = E\left[\left(K - S_T\right)^+\right],$$

the price of a European call respectively put with underlying price process S, strike K and maturity T. These expectations are taken with respect to the initial martingale measure P. If the expectation is to be taken under the dual measure P' we denote the expectation operator by E' and the corresponding prices by $\mathbb{C}'_T\left(S; K\right)$ and $\mathbb{P}'_T\left(S; K\right)$.

Proposition 5.6. *Call-put duality.* *We have*

$$\frac{1}{K}\mathbb{C}_T\left(S; K\right) = \mathbb{P}'_T\left(S'; \frac{1}{K}\right).$$

Proof. Since the density of P' with respect to P is S_T, we get by changing the measure

$$\mathbb{C}_T\left(S; K\right) = E'\left[\frac{\left(S_T - K\right)^+}{S_T}\right] = E'\left[\left(1 - KS'_T\right)^+\right]$$

$$= KE'\left[\left(\frac{1}{K} - S'_T\right)^+\right] = K\mathbb{P}'_T\left(S'; \frac{1}{K}\right).$$

$\square$

Next we consider floating strike **lookback options**. These are call options on a multiple of the infimum of the price process up to maturity. However, the following duality relation is only valid under a symmetry assumption on $X' = -X$.

Definition 5.7. X' satisfies the **dual reflection principle** if

$$\sup_{t \leq T} X'_t - X'_T \sim -\inf_{t \leq T} X'_t,$$

where $\sim$ denotes equality in law under the dual measure P'.

This property is satisfied when X' is a Lévy process, see Kyprianou (2006), Lemma 3.5.

Proposition 5.8. Duality relation for lookback options. *Let X' satisfy the dual reflection principle. Then the price of a lookback call with floating strike equals the price of a lookback put with fixed strike. Specifically, for $\alpha \geq 1$,*

$$\mathbb{C}_T\left(S; \alpha \inf S\right) = \alpha \mathbb{P}'_T\left(\inf S'; \frac{1}{\alpha}\right).$$

Proof. By changing the measure,

$$\mathbb{C}_T\left(S; \alpha \inf S\right) = E\left[\left(S_T - \alpha \inf_{t \leq T} S_t\right)^+\right]$$

$$= E'\left[\left(1 - \alpha \exp\left(\inf_{t \leq T} X_t - X_T\right)\right)^+\right]$$

$$= \alpha E'\left[\left(\frac{1}{\alpha} - \exp\left(X'_T - \sup_{t \leq T} X'_t\right)\right)^+\right]$$

$$= \alpha E'\left[\left(\frac{1}{\alpha} - \exp\left(\inf_{t \leq T} X'_T\right)\right)^+\right],$$

where the last equality follows by the dual reflection principle. $\qquad\square$

Turning our attention to floating strike **Asian options**, the duality principle allows to replace them with simpler fixed strike Asian options, given the following property is satisfied.

Definition 5.9. X' satisfies the **time inversion principle** if

$$X'_T - X'_{(T-t)-} \sim X'_t$$

where $\sim$ denotes equality in law under P'.

This principle holds for Lévy processes X', see Kyprianou (2006), Lemma 3.4.

Proposition 5.10. Duality relation for Asian options. *Let X' satisfy the time inversion principle. Then the price of an Asian option with floating strike equals the price of an Asian option with fixed strike. More precisely,*

$$\mathbb{C}_T\left(S; \frac{1}{T}\int S\,dt\right) = \mathbb{P}'_T\left(\frac{1}{T}\int S'\,dt; 1\right).$$

Proof. We have

$$
\mathbb{C}_T\left(S; \frac{1}{T}\int S\, dt\right) = E\left[\left(S_T - \frac{1}{T}\int_0^T S_t\, dt\right)^+\right]
$$

$$
= E'\left[\left(1 - \frac{1}{T}\int_0^T \frac{S_t}{S_T}\, dt\right)^+\right]
$$

$$
= E'\left[\left(1 - \frac{1}{T}\int_0^T e^{X_T' - X_t'}\, dt\right)^+\right]
$$

$$
= E'\left[\left(1 - \frac{1}{T}\int_0^T e^{X_T' - X_{(T-u)-}'}\, du\right)^+\right].
$$

By the time inversion principle, the last term equals

$$
E'\left[\left(1 - \frac{1}{T}\int_0^T e^{X_u'}\, du\right)^+\right] = E'\left[\left(1 - \frac{1}{T}\int_0^T S_u'\, du\right)^+\right]
$$

which concludes the proof. $\square$

A **Margrabe option** is a call option on the spread between two assets $S^1 = \exp(X^1)$, $S^2 = \exp(X^2)$. Its payoff is $\left(S_T^1 - S_T^2\right)^+$. In general, $\widehat{S} = \exp(X^2 - X^1)$ is not a P'-martingale. If $\widehat{S}$ is a special semi-martingale, then by Jacod & Shiryaev (2003), Theorem II.8.21, it has a multiplicative decomposition as

$$
\widehat{S} = \overline{S}e^C, \tag{5.5}
$$

where $\overline{S}$ is a positive local P'-martingale and e^C a predictable finite variation process. If X^1, X^2 are Lévy processes, then C can be chosen to be deterministic.

Proposition 5.11. *Duality principle for Margrabe options.* *Let $S^1 = \exp(X^1)$ be a martingale, $X_0^1 = 0$, and let $S^2 = \exp(X^2)$, $X_0^2 = 0$, be such that $\widehat{S} = \exp(X^2 - X^1)$ is a special semi-martingale. Assume that C in (5.5) is deterministic. Then the price of the Margrabe option can be expressed in terms of the price of a standard put option,*

$$
\mathbb{C}_T\left(S^1; S_T^2\right) = e^{C_T}\mathbb{P}_T'\left(\overline{S}; e^{-C_T}\right).
$$

Proof. By changing to the dual measure P' (with respect to S^1),

$$\mathbb{C}_T\left(S^1;S_T^2\right) = E\left[\left(S_T^1 - S_T^2\right)^+\right] = E'\left[\left(1 - \frac{S_T^2}{S_T^1}\right)^+\right]$$

$$= E'[(1 - \widehat{S}_T)^+] = E'\left[\left(1 - \overline{S}_T e^{C_T}\right)^+\right]$$

$$= e^{C_T} E'\left[\left(e^{-C_T} - \overline{S}_T\right)^+\right]. \qquad \square$$

5.3 Symmetry and self-dual processes

5.3.1 *Definitions and general properties*

Definition 5.12. Let M be an adapted process. M is **symmetric** if for any stopping time $\tau \in [0,T]$ and any non-negative Borel function f

$$E\left[f\left(M_T - M_\tau\right)|\,\mathcal{F}_\tau\right] = E\left[f\left(M_\tau - M_T\right)|\,\mathcal{F}_\tau\right]. \qquad (5.6)$$

Here it is permissible that both sides of the equation equal $+\infty$. If M is an integrable symmetric process, then condition (5.6) implies by choosing $f(x) = x\ (= x^+ - x^-)$ that M is a martingale.

Definition 5.13. Let S be a positive adapted process. S is **self-dual** if for any stopping time $\tau \in [0,T]$ and any non-negative Borel function f we have

$$E\left[f\left(\frac{S_T}{S_\tau}\right)\Big|\,\mathcal{F}_\tau\right] = E\left[\frac{S_T}{S_\tau}f\left(\frac{S_\tau}{S_T}\right)\Big|\,\mathcal{F}_\tau\right]. \qquad (5.7)$$

In the case when S is a martingale, we define a probability measure P' via

$$\frac{dP'}{dP} = \frac{S_T}{S_0}. \qquad (5.8)$$

Similarly, if $E\left[\sqrt{S_T}\right] < \infty$ we define a probability measure H, sometimes called 'half measure', via

$$\frac{dH}{dP} = \frac{\sqrt{S_T}}{E\left[\sqrt{S_T}\right]}. \qquad (5.9)$$

By Bayes' formula, the self-duality condition (5.7) can be expressed in terms of the measure P' defined in (5.8) as

$$E_P\left[f\left(\frac{S_T}{S_\tau}\right)\Big|\,\mathcal{F}_\tau\right] = E_{P'}\left[f\left(\frac{S_\tau}{S_T}\right)\Big|\,\mathcal{F}_\tau\right].$$

By considering moment generating functions one derives the following lemma, see Tehranchi (2009), Lemma 3.2.

Lemma 5.14. *A martingale S is self-dual if and only if*

$$E\left[\left(\frac{S_T}{S_\tau}\right)^p \middle| \mathcal{F}_\tau\right] = E\left[\left(\frac{S_T}{S_\tau}\right)^{1-p} \middle| \mathcal{F}_\tau\right] \tag{5.10}$$

for all complex $p = a + ib$ with $a \in [0,1]$ and all stopping times $\tau \in [0,T]$.

Proposition 5.15. *Relation between symmetry under H and self-duality.* *Let $S = \exp(X)$ be a martingale, then S is self-dual if and only if X is symmetric with respect to H.*

Proof. Let S be self-dual, then Bayes' formula and (5.10) imply that for $\theta \in \mathbb{R}$

$$E_H\left[e^{i\theta(X_T - X_\tau)} \middle| \mathcal{F}_\tau\right] = \frac{E_P\left[e^{(1/2 + i\theta)(X_T - X_\tau)} \middle| \mathcal{F}_\tau\right]}{E_P\left[e^{1/2(X_T - X_\tau)} \middle| \mathcal{F}_\tau\right]}$$

$$= \frac{E_P\left[e^{(1/2 - i\theta)(X_T - X_\tau)} \middle| \mathcal{F}_\tau\right]}{E_P\left[e^{1/2(X_T - X_\tau)} \middle| \mathcal{F}_\tau\right]}$$

$$= E_H\left[e^{-i\theta(X_T - X_\tau)} \middle| \mathcal{F}_\tau\right].$$

Since the conditional characteristic functions of $(X_T - X_\tau)$ and $(X_\tau - X_T)$ under H are equal, it follows that X is H-symmetric.

On the other hand, if X is H-symmetric, then for $p = a + ib$ with $a \in [0,1]$ Bayes' formula yields that

$$E_P\left[\left(\frac{S_T}{S_\tau}\right)^p \middle| \mathcal{F}_\tau\right] = E_H\left[e^{(p-1/2)(X_T - X_\tau)} \middle| \mathcal{F}_\tau\right] E_P\left[e^{(X_T - X_\tau)/2} \middle| \mathcal{F}_\tau\right]$$

$$= E_H\left[e^{(1/2 - p)(X_T - X_\tau)} \middle| \mathcal{F}_\tau\right] E_P\left[e^{(X_T - X_\tau)/2} \middle| \mathcal{F}_\tau\right]$$

$$= E_P\left[\left(\frac{S_T}{S_\tau}\right)^{1-p} \middle| \mathcal{F}_\tau\right].$$

The statement then follows by Lemma 5.14. $\qquad\square$

Definition 5.16. An adapted positive process S is **quasi self-dual** of order $\alpha \neq 0$ if for any stopping time τ and any non-negative Borel function f it holds that

$$E_P\left[f\left(\frac{S_T}{S_\tau}\right) \middle| \mathcal{F}_\tau\right] = E_P\left[\left(\frac{S_T}{S_\tau}\right)^\alpha f\left(\frac{S_\tau}{S_T}\right) \middle| \mathcal{F}_\tau\right].$$

Proposition 5.17. *Characterisation of quasi self-duality.* *S is quasi self-dual if and only if there exists $\alpha \neq 0$ such that S^α is self-dual.*

Proof. This follows by considering for each f the functions g defined by $g(x) = f(x^\alpha)$, respectively h given by $h(x) = f(x^{1/\alpha})$, $x > 0$. □

5.3.2 *Semi-static hedging of barrier options*

Let S be a positive martingale modelling the asset price process under some risk-neutral measure. We fix a barrier $H \neq S_0$, let $\eta := sgn(H - S_0)$ and denote by $\tau_H := \inf\{t \,|\, \eta S_t \geq \eta H\}$ the hitting time of the barrier. The indicator function of the event that the barrier is hit before the maturity T is denoted by $\chi := \mathbb{I}_{\tau_H \leq T}$. We consider a payoff g such that $g(S_T)$ is integrable, and the payoff of the corresponding knock-in barrier option is $\chi g(S_T)$. Hence once the barrier has been hit, the barrier option is equivalent to a European option. We aim to describe a semi-static hedging strategy to replicate the barrier which involves trading in European options at most twice. The first trade occurs at time zero, and in case $\tau_H \leq T$ the second takes place at the hitting time of the barrier.

Proposition 5.18. *Replication of barrier options.* *Assume that S is quasi self-dual of order α. Moreover, assume $S_{\tau_H} = H$ if $\tau_H \leq T$, and $P(S_T = H) = 0$. Then the following strategy provides a semi-static hedge for the knock-in claim $\chi g(S_T)$:*

(1) At time zero, purchase and hold a European claim

$$G(S_T) = \left(g(S_T) + \left(\frac{S_T}{H}\right)^\alpha g\left(\frac{H^2}{S_T}\right)\right) \mathbb{I}_{\eta S_T > \eta H}.$$

(2) If and when the barrier knocks in, exchange the $G(S_T)$-claim for a $g(S_T)$-claim, at zero cost.

Proof. In case $\tau_H > T$, both the barrier option and the claim $G(S_T)$ expire worthless. Let now $\tau_H \leq T$. Then, by quasi self-duality we have

$E[g(S_T)|\mathcal{F}_{\tau_H}]$

$= E[g(S_T)\mathbb{I}_{\eta S_T > \eta H}|\mathcal{F}_{\tau_H}] + E[g(S_T)\mathbb{I}_{\eta S_T < \eta H}|\mathcal{F}_{\tau_H}]$

$= E[g(S_T)\mathbb{I}_{\eta S_T > \eta H}|\mathcal{F}_{\tau_H}] + E\left[\left(\frac{S_T}{H}\right)^\alpha g\left(\frac{H^2}{S_T}\right)\mathbb{I}_{\eta H^2/S_T < \eta H}\middle|\mathcal{F}_{\tau_H}\right]$

$= E\left[\left(g(S_T) + \left(\frac{S_T}{H}\right)^\alpha g\left(\frac{H^2}{S_T}\right)\right)\mathbb{I}_{\eta S_T > \eta H}\middle|\mathcal{F}_{\tau_H}\right].$

□

(A) If S is a geometric Brownian motion with constant carrying costs r,

$$dS = rS\,dt + \sigma S\,dW,$$

then Theorem 5.25 shows that $e^{-rt}S$ is quasi self-dual of order $\alpha = 1 - 2r/\sigma^2$. The assumptions of Proposition 5.18 are fulfilled since S has continuous paths and the distribution of S_T is atomless.

(B) When S is an exponential Lévy process, the assumptions of Proposition 5.18 may be violated if S has positive jumps. However, when $\eta = +1$ (respectively $\eta = -1$), and $G - g$ is an increasing (respectively decreasing) function, then it follows from the strong Markov property and $S_{\tau_H} \geq H$ that there exist an increasing (decreasing) function ϕ such that

$$E\left[G(S_T) - g(S_T)|\, \mathcal{F}_{\tau_H}\right] = \phi\left(S_{\tau_H}\right) \geq \phi\left(H\right) = 0.$$

Therefore, in this case the hedging portfolio described in Proposition 5.18 super-replicates the knock-in option.

Remark. In practice, the claim $G(S_T)$ could be synthetically approximated using vanilla options as in (5.4).

5.3.3 *Self-dual exponential Lévy processes*

Proposition 5.19. *Self-duality for exponential Lévy processes.* *Let* $S = \exp(X)$ *be a martingale with Lévy process* X, $X_0 = 0$, *which satisfies* (EM) *and has Lévy measure* ν. *Then* S *is self-dual if and only if*

$$\nu\left(dx\right) = e^{-x}\,\nu\left(-dx\right). \tag{5.11}$$

Proof. Since self-duality implies that the Lévy measure ν of X with respect to P equals the dual Lévy measure ν' of $-X$ with respect to P', the first implication follows immediately from Proposition 5.4. Conversely, as $\exp\left(X_T\right)$ is integrable, the H-measure, see (5.9), is well-defined. The H-measure is an Esscher transform with parameter $\theta = 1/2$, and the Lévy triplet $\left(\gamma^H, \sigma^2, \nu^H\right)$ of X under H can be computed as in (9.4). (5.11) implies then that ν^H is a symmetric measure. This, together with the martingale condition (4.6) yields that $\gamma^H = 0$. Therefore, X is H-symmetric, and hence self-dual by Proposition 5.15. $\square$

Theorem 5.20. *Quasi self-duality for exponential Lévy processes.* *Let* $S = e^{\lambda t}\exp(X)$ *for* $\lambda \in \mathbb{R}$ *and a Lévy process* X, $X_0 = 0$, *which satisfies* (EM) *and has triplet* $\left(\gamma, \sigma^2, \nu\right)$. *Assume that* $\exp(X)$ *is a martingale, and* *that* $\exp(\alpha X)$ *is integrable for* $\alpha \neq 0$. *Then* S *is quasi self-dual of order* α *if and only if the following conditions hold.*

(i) *The Lévy measure satisfies*

$$\nu\left(dx\right) = e^{-\alpha x}\,\nu\left(-dx\right). \tag{5.12}$$

(ii) *The entries of the triplet satisfy the* H*-martingale condition*

$$\gamma = \int x\left(1 - e^{\frac{\alpha}{2}x}\right)\nu\left(dx\right) - \frac{1}{2}\alpha\sigma^2 - \lambda. \tag{5.13}$$

(iii) *The parameters* λ *and* α *are related by*

$$\lambda = (1-\alpha)\frac{\sigma^2}{2} + \int\left(e^x - xe^{\frac{\alpha}{2}x} - 1\right)\nu\left(dx\right). \tag{5.14}$$

Proof. *ad (i):* By Proposition 5.17, S quasi self-dual implies that $S^\alpha = e^{\alpha\lambda t}\exp\left(\alpha X\right)$ is self-dual for some $\alpha \neq 0$. Denoting by $\widehat{\nu}$ the Lévy measure of $\widehat{X} = \alpha X$, we therefore have by Proposition 5.19 that

$$\nu\left(dx\right) = \widehat{\nu}\left(\alpha\,dx\right) = e^{-\alpha x}\widehat{\nu}\left(-\alpha\,dx\right) = e^{-\alpha x}\nu\left(-dx\right).$$

ad (ii): According to Proposition 5.15, $\alpha\lambda t + \alpha X$ is a H-martingale. The measure H is an Esscher transform with parameter $\alpha/2$, and we calculate the triplet of X under H by Theorem 4.21. The H-martingale condition (4.6) then yields (5.13).

ad (iii): Equation (5.14) follows from the H-martingale condition (5.13) together with the P-martingale condition (4.6) for $\exp(X)$,

$$\gamma = -\frac{\sigma^2}{2} + \int\left(x + 1 - e^x\right)\nu\left(dx\right).$$

Conversely, conditions (i) and (ii) imply by considering the Esscher transform with parameter $\alpha/2$ the quasi self-duality of S along similar lines as in the proof of Proposition 5.19. □

Remark. Equation (5.14) has been studied for a variety of specific exponential Lévy models in Rheinländer & Schmutz (2011).

5.3.4 *Self-dual stochastic volatility models*

In the sequel, we assume that Y is a continuous martingale with $Y_0 = 0$. Let $X = Y - \frac{1}{2}[Y]$ and observe that $[X] = [Y]$, hence $Y = X + \frac{1}{2}[X]$, so that $S = \exp(X) = \mathcal{E}(Y)$. The next result is more difficult to prove than its counterpart Proposition 5.15, and we refer to Tehranchi (2009), Theorem 3.1.

Theorem 5.21. *Relation between symmetry under P and self-duality.* *Let Y be a continuous P-martingale and assume that $S = \mathcal{E}(Y)$ is a martingale. Then Y is symmetric if and only if S is self-dual.*

In contrast to Y, the process $X = \log(S)$ is typically not a martingale. As $X = Y - \frac{1}{2}[Y]$, the minimal martingale measure (see Section 9.5) for X is well-defined if $\mathcal{E}\left(\frac{1}{2}Y\right)$ is a martingale, and has the density

$$\frac{d\widehat{P}}{dP} = \exp\left(\frac{1}{2}Y_T - \frac{1}{8}[Y]_T\right) = \exp\left(\frac{1}{2}X_T + \frac{1}{8}[X]_T\right).$$

In Section 7.2 we will discuss the minimal entropy martingale measure Q^0 for X which can be characterised as the martingale measure for X such that

$$\frac{dQ^0}{dP} = \exp\left(c + \int_0^T \eta_t\, dX_t\right),$$

where η is a predictable process with the property that $\int \eta\, dX$ is a Q-martingale for all martingale measures Q with finite relative entropy, see Theorem 7.14. We have seen in Proposition 5.15 that under mild conditions the measure H with density

$$\frac{dH}{dP} = \exp\left(c + \frac{1}{2}X_T\right)$$

is a martingale measure for X. The preceding discussion shows that typically, H is the minimal entropy martingale measure and $\eta = 1/2$. This is different from the minimal martingale measure in the case of orthogonal volatility models, see Section 9.2.4. Moreover, it is remarkable that $Q^0 = H$ has such a simple form, which has consequences for the structure of symmetric martingales.

Theorem 5.22. *Self-duality and PRP.* *Let Y be a symmetric continuous P-martingale which has the PRP. Assume that $S = \mathcal{E}(Y)$ is a martingale, and that the minimal martingale measure $\widehat{P}$ exists for $X = Y - \frac{1}{2}[Y]$. Then Y is a Gaussian martingale.*

Proof.   The fact that Y has the PRP under P implies that X has the PRP under $\widehat{P}$. Moreover, existence of $\widehat{P}$ implies existence of the probability measure H as defined in (5.9) because

$$E\left[\exp\left(\frac{1}{2}X_T\right)\right] \leq E\left[\exp\left(\frac{1}{2}X_T + \frac{1}{8}[X]_T\right)\right] = E\left[\frac{d\widehat{P}}{dP}\right] = 1.$$

Since Y is symmetric, it follows from Theorem 5.21 that S is self-dual, and hence by Proposition 5.15 X is a symmetric martingale under H. In particular, H is a martingale measure for X. The PRP implies by the second fundamental theorem of asset pricing that $\widehat{P} = H$ which yields

$$\exp\left(\frac{1}{2}Y_T - \frac{1}{8}[Y]_T\right) = \exp\left(\frac{1}{2}X_T + \frac{1}{8}[X]_T\right) = \exp\left(c + \frac{1}{2}X_T\right).$$

It follows that $[Y] = [X]$ must be deterministic, and therefore Y is a Gaussian martingale. □

A family of martingales which are symmetric are the Ocone martingales, see Section 4.2.2 for their definition. They can be characterised as follows: a martingale M is Ocone if and only if $\int \left(\mathbb{I}_{[0,s]} - \mathbb{I}_{(s,\infty)}\right) dM_s$ and M have the same law for every $s \geq 0$; see Ocone (1993). The following result explores the relationship between continuous symmetric and Ocone martingales, via the link to self-dual stochastic exponentials.

Theorem 5.23. *Self-duality and Ocone martingales.* *Let Y be a continuous martingale. Then $\mathcal{E}(pY)$ is self-dual and a martingale for all $p \in \mathbb{R}$ if and only if Y is an Ocone martingale.*

Proof.   If Y is Ocone, with DDS representation $Y = B_{[Y]}$ for a Brownian motion B independent of $[Y]$, then $\mathcal{E}(pY)$ is a martingale for all $p \in \mathbb{R}$ which follows readily by conditioning on $\mathcal{F}^{[Y]}_T$. Let $\mathcal{E}(pY) = \exp(X^{(p)})$ where $X^{(p)} = pY - \frac{1}{2}p^2[Y]$, and define probability measures $Q^{(p)}$ via $dQ^{(p)}/dP = \mathcal{E}(pY)_T$. We show that the symmetries of the $X^{(p)}$ under the probability measures $H^{(p)}$ with densities $dH^{(p)}/dP = \exp\left(\frac{1}{2}X^{(p)}_T\right)/c_p$, for all $p \in \mathbb{R}$, are equivalent to the Ocone property of Y. This yields the result by Proposition 5.15.

It follows from the proof of Theorem 1 of Vostrikova & Yor (2000) that Y is Ocone if and only if for each $p \in \mathbb{R}$, $Y - p[Y]$ under $Q^{(p)}$ has the same law as Y under P. Note that $[X^{(p)}] = p^2[Y]$, and therefore

$Y - p[Y] = \frac{1}{p}\left(X^{(p)} - \frac{1}{2}\left[X^{(p)}\right]\right)$ for $p \neq 0$. We have for $\theta \in \mathbb{R}$

$$E_{Q^{(p)}}[\exp(i\theta(Y_T - p[Y]_T))]$$

$$= E_P\left[\exp\left(pY_T - \frac{p^2}{2}[Y]_T + i\theta\left(Y_T - p[Y]_T\right)\right)\right]$$

$$= E_P\left[\exp\left(X_T^{(p)} + i\frac{\theta}{p}\left(X_T^{(p)} - \frac{1}{2}\left[X^{(p)}\right]_T\right)\right)\right]$$

$$= c_p E_{H^{(p)}}\left[\exp\left(\frac{1}{2}X_T^{(p)} + i\frac{\theta}{p}\left(X_T^{(p)} - \frac{1}{2}\left[X^{(p)}\right]_T\right)\right)\right].$$

On the other hand,

$$E_P\left[\exp\left(-i\theta Y_T\right)\right] = E_P\left[\exp\left(-i\frac{\theta}{p}\left(X_T^{(p)} + \frac{1}{2}\left[X^{(p)}\right]_T\right)\right)\right]$$

$$= c_p E_{H^{(p)}}\left[\exp\left(-\frac{1}{2}X_T^{(p)} + i\frac{\theta}{p}\left(-X_T^{(p)} - \frac{1}{2}\left[X^{(p)}\right]_T\right)\right)\right].$$

These equalities are readily extended to the conditional case using Bayes' formula. Hence the symmetry of $X^{(p)}$ under $H^{(p)}$ is equivalent to that $Y - p[Y]$ under $Q^{(p)}$ has the same law as $-Y$ under P. The statement follows since by Proposition 5.15 and Theorem 5.21, self-duality of $S^{(p)}$ is equivalent to both symmetry of $X^{(p)}$ under $H^{(p)}$, and pY, hence Y, under P. □

Let us now discuss some examples of symmetric SV models.

Proposition 5.24. *Symmetric SV models.* *Let (Y,V) be a weak solution of the SDE*

$$dY_t = \sigma\left(Y_t, V_t\right) dB_t, \tag{5.15}$$

$$dV_t = \alpha\left(Y_t, V_t\right) dt + \beta\left(Y_t, V_t\right) dW_t,$$

where B, W are two uncorrelated Brownian motions, and $\sigma\left(y, v\right)$, $\alpha\left(y, v\right)$, $\beta\left(y, v\right)$ are even functions in y such that uniqueness in law holds for equation (5.15), and that Y is a martingale with $Y_0 = 0$. Then Y is symmetric, and $S = \mathcal{E}\left(Y\right)$ self-dual.

Proof. Since $-B$ is a Brownian motion, the assumptions imply that

$$d\left(-Y\right)_t = \sigma\left(\left(-Y\right)_t, V_t\right) d\left(-B\right)_t,$$

$$dV_t = \alpha\left(\left(-Y\right)_t, V_t\right) dt + \beta\left(\left(-Y\right)_t, V_t\right) dW_t.$$

Hence, both $((Y,V);(B,W))$ and $((-Y,V);(-B,W))$ are weak solutions to (5.15). By uniqueness in law, the conditional distribution of $Y_T - Y_\tau$ is therefore the same as that of $-(Y_T - Y_\tau)$. The self-duality of S then follows by Theorem 5.21. □

(A) Consider as special case of (5.15), with the same assumptions in place, the SDE

$$dY_t = V_t \, dB_t,$$
$$dV_t = \alpha \, (V_t) \, dt + \beta \, (V_t) \, dW_t.$$

Then X is symmetric with respect to the filtration generated by B and W, and in fact, an Ocone martingale.

(B) The martingales $\int B^{2n} \, dB$, $n \in \mathbb{N}$, are symmetric with respect to the filtration generated by B, since

$$-\int B^{2n} \, dB = \int (-B)^{2n} \, d(-B).$$

Moreover, they have the PRP, but are non-Gaussian, and hence not Ocone. More interestingly, the results in Beghdadi-Sakrani (2003) show that these martingales are even symmetric with respect to their own filtration which is shown to equal the one generated by B. In particular, it is demonstrated that these martingales are pure.

The next result completely characterises quasi self-dual continuous processes. While the notion of quasi self-duality seems capable to deal with the costs of carry as captured by the term $e^{\lambda[M]}$, essentially S is still linked to the stochastic exponential of a symmetric process.

Theorem 5.25. *A continuous positive process S is quasi self-dual of order $\alpha = 1 - 2\lambda$, $\alpha \neq 0$, if and only if S^α is a martingale and $S = e^{\lambda[M]}\mathcal{E}(M)$ for a continuous symmetric martingale M.*

Proof. S is quasi self-dual if and only if S^α is self-dual for some $\alpha \neq 0$, and hence in particular S^α is a positive continuous martingale. We can then write by Theorem 5.21

$$S^\alpha = \mathcal{E}(\alpha M) = \exp\left(\alpha M - \frac{1}{2}\alpha^2 [M]\right) = e^{\alpha\lambda[M]}\mathcal{E}(M)^\alpha$$

for some symmetric martingale M. On the other hand, if $S = e^{\lambda[M]}\mathcal{E}(M)$, we have

$$S^\alpha = e^{\alpha\lambda[M]}\mathcal{E}(M)^\alpha = \exp\left(\alpha M + \left(\lambda - \frac{1}{2}\right)\alpha [M]\right).$$

It follows that S^α is a martingale if and only if $S^\alpha = \mathcal{E}(\alpha M)$ which is equivalent to

$$\left(\lambda - \frac{1}{2}\right)\alpha = -\frac{1}{2}\alpha^2. \qquad \square$$

Remark. If M is both a continuous martingale and a Lévy process, then the parameter λ in Proposition 5.25 is, with a slight abuse of notation, different from λ as in Theorem 5.20. This is due to the fact that we are here working on the time scale $[M]$ as opposed to t.

5.4 Notes and further reading

The Taylor representation for European claims has been generalised to payoff functions which can be represented as the difference of two convex functions whose right derivatives in 0 are finite, and where the integrals are interpreted in the Lebesgue-Stieltjes sense, see Schmutz & Zürcher (2011). Moreover, there exists a multivariate extension by Schmutz & Zürcher (2010) where, in addition to calls, puts and forwards, one needs to invest in so-called traffic light options to build up the static hedge.

The duality principle for general semi-martingales has been studied by Eberlein *et al.* (2008), and the duality relations in a Lévy process setting have been obtained by Eberlein & Papapantoleon (2005). Semi-static hedging of barrier options and self-duality is discussed in Carr & Lee (2009), as well as in Molchanov & Schmutz (2010) for the multivariate case. Symmetric continuous martingales is the subject matter of Tehranchi (2009). For quasi self-dual processes, see Rheinländer & Schmutz (2011).

Chapter 6

Mean-Variance Hedging

In an incomplete market, not every claim is replicable by trading with the underlying asset. In this chapter, we aim to minimise the remaining risk using a quadratic criterion, formulated under a martingale measure. A quadratic criterion can be considered under the statistical measure as well, but the resulting theory is considerably more difficult than the one discussed here. We will review mean-variance hedging under the statistical measure in Section 9.3.

At first glance, the choice of a quadratic criterion may seem inappropriate as gains and losses are punished symmetrically. However, the quadratic criterion is imposed only after we have switched from the statistical to a risk-neutral measure. Hence, as seen under the statistical measure, this approach need not be symmetric. In fact, empirical studies suggest that changing to the risk-neutral world introduces an asymmetry when it comes to option pricing; see Cont & Tankov (2004), (2006).

6.1 Concept of mean-variance hedging

To prepare for the following discussion, the reader is encouraged to revise notation and results from Section 2.3. We assume:

> The price process S is a locally square-integrable P-martingale.

For the purpose of mean-variance hedging, the following strategy set is appropriate:

Definition 6.1. A strategy ϑ is called **admissible** if $\vartheta \in L^2(S)$.

Note that although S might be a strict local martingale, the gains process $\int \vartheta \, dS$ is a square-integrable martingale for every admissible strategy ϑ. Given a claim $H \in L^2(\mathcal{F}_T, P)$, define a martingale $V \in \mathcal{M}^2$ via

$$V_t := E\left[H \mid \mathcal{F}_t\right], \qquad 0 \le t \le T. \tag{6.1}$$

For an initial capital c and a self-financing strategy ϑ, the associated value process is $c + \int \vartheta \, dS$. Our goal is to minimise

$$E\left[\left(H - c - \int_0^T \vartheta_t \, dS_t\right)^2\right] \tag{6.2}$$

over all constants c and all $\vartheta \in L^2(S)$. By identifying square-integrable random variables as terminal values of martingales in $\mathcal{M}^2$ (like in (6.1)), minimising (6.2) corresponds to the orthogonal projection of $V \in \mathcal{M}^2$ onto the closed subspace spanned by the constants and the stable subspace $\mathcal{S}(S)$. Therefore the next result is an immediate consequence of the orthogonal projection theorem for Hilbert spaces.

Theorem 6.2. *Optimal mean-variance hedging strategy.* *Consider the KW decomposition of V,*

$$V = E\left[H\right] + \int \vartheta^H \, dS + L, \tag{6.3}$$

with $\vartheta^H \in L^2(S)$ and $L \in \mathcal{M}_0^2$, strongly orthogonal to each element of $\mathcal{S}(S)$ (and hence a fortiori orthogonal in the sense of $\mathcal{M}^2$). The optimal initial capital c^* and optimal strategy ϑ^* minimising the quadratic functional (6.2) are $c^* = E\left[H\right]$, $\vartheta^* = \vartheta^H$. The optimal strategy is unique in the sense that for two optimal strategies ϑ^*, ψ^* the resulting stochastic integral processes are indistinguishable, or equivalently, $\int \left(\vartheta^* - \psi^*\right)^2 d\left[S\right] = 0$.

One can interpret $c^* + \int \vartheta^H \, dS$ as the part of the risk which is attainable, so can be perfectly replicated by means of the hedging strategy ϑ^H, whereas, L is the part of the risk that is totally unhedgeable. Thus L_T is the risk-component of the claim H that cannot be accessed by trading in the underlying. To quantify this inaccessible risk, we are often interested in calculating the variance of the remaining hedging error,

$$R\left(\vartheta^H\right) := E\left[L_T^2\right].$$

Since by strong orthogonality (6.3) implies that $\langle V, S \rangle = \int \vartheta^H \, d\langle S, S \rangle$, we can determine the optimal mean-variance hedging strategy ϑ^H by calculating the formal derivative $\vartheta^H = d\langle V, S \rangle / d\langle S, S \rangle$.

To sum up, the mean-variance hedging approach is a method yielding both a fair price c^* and an optimal hedging strategy ϑ^*. In particular, the fair price is the expectation of the claim under the chosen martingale measure which seems to be a very suitable extension of the pricing rule for complete markets.

The purpose of the next two sections is to apply the general method of this section to options written on price processes which are given as exponential Lévy processes, $S = S_0 \exp(X)$. In particular, there are two popular approaches: the first exploits that often the characteristic function of X_T is of simpler structure than the distribution by considering Laplace transforms; the second uses a Feynman-Kac-type argument to relate the value process of a claim to the solution of an integro-partial differential equation. We then discuss mean-variance hedging of claims on defaultable assets, employing the reduced-form approach to credit risk.

In the framework discussed so far, one may insist on perfect replication of the claim, but for non-attainable claims this is only possible when using strategies which are no longer self-financing. Indeed, perfect replication is achieved by the portfolio which at time t consists of ϑ_t^H shares and L_t units in the bank account. The value process at time T of this generalised strategy (ϑ^H, L) equals then by (6.3) together with $V_T^H = H$ the payoff of the claim. The cost of this portfolio is given by the zero-mean martingale L and in this sense the strategy (ϑ^H, L) is mean self-financing. This approach will be studied in the last section of this chapter, in the more general context of payment streams.

6.2 Valuation and hedging by the Laplace method

6.2.1 *Bilateral Laplace transforms*

Definition 6.3. For a Borel measurable function $g : \mathbb{R} \to \mathbb{C}$, its **bilateral Laplace transform** $\mathcal{L}_g : \mathbb{C} \to \mathbb{C}$ is defined as

$$\mathcal{L}_g(z) = \int_{-\infty}^{+\infty} g(x) e^{-zx} \, dx \qquad (6.4)$$

for any $z \in \mathbb{C}$ such that the integral exists (as Lebesgue integral).

Henceforth, we will often skip the qualifier 'bilateral' and refer to the function $\mathcal{L}_g$ defined above as the Laplace transform. Note that if the integral in (6.4) exists for real numbers $-\infty \le a \le b \le +\infty$, it also exists for any $z \in \mathbb{C}$ with $a \leq \Re(z) \leq b$. This follows from the estimate

$$\left|g(x)e^{-zx}\right| = |g(x)| \, e^{-\Re(z)x} \leq \left|g(x)e^{-ax}\right| + \left|g(x)e^{-bx}\right|.$$

Let X be a random variable admitting a density ρ, and let ϕ be its characteristic function. Then, for all $z \in \mathbb{C}$ in the domain of the Laplace transform,

$$\mathcal{L}_\rho(z) = E\left[\exp\left(-zX\right)\right] = \phi(iz), \tag{6.5}$$

and hence the Laplace transform can be expressed in terms of the characteristic function ϕ. The Laplace transform $\mathcal{L}_g$ determines the function g uniquely: if $v \mapsto g(R + iv)$ is integrable, then

$$g(\zeta) = \frac{1}{2\pi i} \int_{R-i\infty}^{R+i\infty} e^{\zeta z} \mathcal{L}_g(z) \, dz \tag{6.6}$$

where the integral is taken along a line parallel to the imaginary axis, with real part R. This integral is known as the **Bromwich inversion integral**. In particular, as long as g satisfies the integrability condition, the value of the integral does not depend on R which can be shown using Cauchy's integral theorem from complex contour integration.

6.2.2   *Valuation and hedging using Laplace transforms*

Let X be a Lévy process and denote by $\mathbb{F}$ its augmented natural filtration. The distribution of the process X is fixed given the distribution of the random variable X_1. By setting

$$D := \left\{ z \in \mathbb{C} \,\middle|\, E\left[e^{\Re(z)X_1}\right] < \infty \right\},$$

we define the cumulant generating function κ for $t > 0$ as

$$\kappa : \; D \to \mathbb{C}$$
$$e^{\kappa(z)t} = E\left[e^{zX_t}\right].$$

To ensure that X is non-deterministic, we exclude that $\kappa(2)$ equals zero. For instance, if $X = \sigma B - \frac{1}{2}\sigma^2 t$ where B is a standard Brownian motion, then $\kappa(2) = \sigma^2$.

We consider now the (discounted) price process $S = S_0 \exp(X)$ and assume that

> S is a square-integrable P-martingale.

As a consequence of this assumption, it follows from the martingale property of $\exp\left(X - \kappa\left(1\right)t\right)$, see Proposition 4.9, that $\kappa\left(1\right) = 0$. Moreover, since S is square-integrable,

$$E\left[S_1^2\right] = S_0^2 \, E\left[e^{2X_1}\right] < \infty,$$

which implies that $z \in D$ whenever $0 \leq \Re\left(z\right) \leq 2$.

To apply the Laplace method for hedging we use an integral representation of the option payoff functions. In particular, we will write the option payoff as a linear superposition of power options. That is, we consider a modified payoff function f which can be written as

$$f(s) = \int s^z \, \Pi\left(dz\right), \quad 0 < s < \infty, \tag{6.7}$$

for some finite complex measure Π on a strip $\left\{z \in \mathbb{C} \,|\, R' \leq \Re\left(z\right) \leq R\right\}$ where both $E\left[e^{2R'X_1}\right] < \infty$ and $E\left[e^{2RX_1}\right] < \infty$. For some generic payoffs this measure Π can be determined explicitly using the Bromwich inversion formula (6.6).

Example 6.4. European call option

We will calculate the Laplace transform for a European call option with payoff $h\left(x\right) = \left(x - K\right)^+$. Without loss of generality we focus on the pricing of a European call option with strike equal to one since, with $y = x - \log K$,

$$h\left(e^x\right) = \left(e^x - K\right)^+ = K\left(e^{x - \log K} - 1\right)^+ = K\left(e^y - 1\right)^+.$$

The modified call payoff is therefore $f(x) = \left(\exp(x) - 1\right)^+$. Suppose $C\left(\zeta; K\right)$ denotes the price of a call as a function of $\zeta = \log S_0$ and strike K. Then, it follows from the above that

$$C\left(\zeta; K\right) = K \, C\left(\zeta - \log K; 1\right).$$

Calculating the Laplace transform of the modified call option payoff involves the Euler Beta function, defined for $\Re\left(z\right) > 0$, $\Re\left(w\right) > 0$ as

$$B\left(z, w\right) = \int_0^1 t^{z-1}\left(1 - t\right)^{w-1} dt$$

and its relation to the Gamma function,

$$B\left(z, w\right) = \frac{\Gamma\left(z\right)\Gamma\left(w\right)}{\Gamma\left(z + w\right)}.$$

The bilateral Laplace transform exists in our case for all $z \in \mathbb{C}$ with $\mathfrak{R}(z) > 1$. For all such z we have, using the substitution $t = e^{-x}$, that

$$\int_{-\infty}^{+\infty} f(x) e^{-zx} dx = \int_0^\infty (e^x - 1) e^{-zx} dx = \int_0^1 \left(\frac{1}{t} - 1 \right) t^z \frac{dt}{t}$$

$$= \int_0^1 t^{z-2} (1 - t) \, dt = B(z - 1, 2)$$

$$= \frac{\Gamma(z - 1) \Gamma(2)}{\Gamma(z + 1)} = \frac{1}{(z - 1) z}.$$

Analogously, the modified put option payoff with strike one is $f(x) = (1 - e^x)^+$ and the Laplace transform (valid for all $z \in \mathbb{C}$ with $\mathfrak{R}(z) < 0$) is the same as for the call, namely $1/(z - 1) z$. More generally, for a power call with payoff $f(x) = \left[(e^x - 1)^+ \right]^n$ for some fixed $n > 0$, the Laplace transform is

$$B(z - n, n + 1)$$

which is valid for all $z \in \mathbb{C}$ with $\mathfrak{R}(z) > n$. This is enough to identify Laplace transforms for power calls with general strikes since if $C^n(\zeta; K)$ denotes the price of a power n call as a function of $\zeta = \log S_0$ and strike K, then

$$C^n(\zeta; K) = K^n C^n(\zeta - \log K; 1).$$

Now that the Laplace transform of the call option's modified payoff function has been calculated for $z \in \mathbb{C}$ with $\mathfrak{R}(z) > 1$ to be

$$\int_{-\infty}^{+\infty} (e^x - 1)^+ e^{-zx} dx = \frac{1}{(z - 1) z},$$

the measure $\Pi(dz)$ in (6.7) can be determined for that case. For arbitrary $R > 1$ and $s > 0$, we get by the inversion formula (6.6) that

$$(s - 1)^+ = \frac{1}{2\pi i} \int_{R-i\infty}^{R+i\infty} s^z \frac{1}{(z - 1) z} dz,$$

and therefore

$$\Pi(dz) = \frac{1}{2\pi i} \frac{dz}{(z - 1) z}.$$

Let us now state the two main results. We will use the following quantities:

$$\alpha\,(y,z) := \kappa(y) + \kappa(z),$$

$$\beta\,(y,z) := \kappa(y+z) - \kappa(y) - \kappa(z) - \frac{(\kappa(y+1)-\kappa(y))(\kappa(z+1)-\kappa(z))}{\kappa(2)}$$

$$\gamma\,(z) := \frac{\kappa\,(z+1) - \kappa\,(z)}{\kappa\,(2)}. \tag{6.8}$$

Theorem 6.5. *Mean-variance value, optimal hedging strategy.* *Let*

$$H_t := \int e^{\kappa(z)(T-t)} S_t^z\, \Pi\,(dz),$$

$$\xi_t := \int \gamma\,(z)\, e^{\kappa(z)(T-t)} S_{t-}^{z-1}\, \Pi\,(dz).$$

Then the mean-variance value of the option $f\,(S_T) \in L^2(P)$ at time t, where f is given by (6.7), and with current price $S_t = s$, is

$$V_t = H_t|_{S_t = s} = \int e^{\kappa(z)(T-t)} s^z\, \Pi\,(dz). \tag{6.9}$$

The optimal mean-variance hedging strategy is given by

$$\vartheta^H = \xi \in L^2(S).$$

Theorem 6.6. *Variance of hedging error.* *We have that*

$$R\left(\vartheta^H\right) := E\left[\left(V_0 + \int_0^T \vartheta_t^H\, dS_t - f\,(S_T)\right)^2\right]$$

equals

$$R\left(\vartheta^H\right) = \int\int R\,(y,z)\, \Pi\,(dy)\, \Pi\,(dz)$$

where

$$R(y,z) := \int_0^T e^{\kappa(y+z)t + \alpha(y,z)(T-t)}\, dt.$$

By Theorem 6.2, the key to the proof of these two results is to determine the KW decomposition of $f(S_T)$ (resp. its associated martingale). Firstly, this will be done for power options of the form $f(S_T) = S_T^z$. These results will then be extended via a stochastic Fubini theorem to get the decomposition for general claims.

We start by determining the canonical decomposition of S^z into its local martingale part $M(z)$ and its predictable finite variation part $A(z)$, and calculate the angle bracket process $\langle M(y), M(z) \rangle$.

Lemma 6.7. *Let $z \in \mathbb{C}$ such that $S_T^z \in L^2(P)$, and denote*

$$N(z)_t := e^{-\kappa(z)t} S_t^z.$$

Then S^z is a special semi-martingale with canonical decomposition

$$S^z = S_0^z + M(z) + A(z),$$

where

$$M(z)_t = \int_0^t e^{\kappa(z)u} \, dN(z)_u,$$

$$A(z)_t = \kappa(z) \int_0^t S_u^z \, du.$$

Proof. By definition of the cumulant generating function, $N(z)$ is a martingale. The decomposition follows then by integration by parts applied to $e^{\kappa(z)t} N(z)$. $\qquad\square$

Lemma 6.8. *Take $y, z \in \mathbb{C}$ such that $S_T^y, S_T^z \in L^2(P)$. Then*

$$\langle M(y), M(z) \rangle = (\kappa(y+z) - \kappa(y) - \kappa(z)) \int S^{y+z} \, du.$$

Proof. We have

$$[M(y), M(z)] = [S^y, S^z]$$

$$= S^{y+z} - S_0^{y+z} - \int S_-^y \, dS^z - \int S_-^z \, dS^y$$

$$= \text{local martingale terms}$$

$$+ A(y+z) - \int S_-^y \, dA(z) - \int S_-^z \, dA(y)$$

$$= \text{local martingale terms}$$

$$+ (\kappa(y+z) - \kappa(y) - \kappa(z)) \int S^{y+z} \, du,$$

from which the result follows by the definition of the angle bracket. $\qquad\square$

After these preparations, we can determine the KW-decomposition of

$$H(z)_t := E\left[S_T^z \middle| \mathcal{F}_t \right] = e^{\kappa(z)(T-t)} S_t^z.$$

Proposition 6.9. *KW decomposition for power options.* *Let $z \in \mathbb{C}$ such that $S_T^z \in L^2(P)$. Then*

$$H(z) = H(z)_0 + \int \xi(z)\, dS + L(z),$$

where

$$\xi(z)_t := \gamma(z)\, e^{\kappa(z)(T-t)} S_{t-}^{z-1},$$

$$L(z)_t := H(z)_t - H(z)_0 - \int_0^t \xi(z)_u\, dS_u.$$

Here $L(z) \in \mathcal{M}_0^2$ with $\langle S, L(z) \rangle = 0$. Moreover, with the notation of (6.8),

$$\langle L(y), L(z) \rangle_t = \beta(y,z) \int_0^t e^{\alpha(y,z)(T-u)} S_u^{y+z}\, du. \qquad (6.10)$$

Proof. Integration by parts and the definition of $A(z)$ yield

$$H(z)_t = H(z)_0 + \int_0^t e^{\kappa(z)(T-u)}\, dM(z)_u.$$

Therefore, by definition of $L(z)$,

$$L(z)_t = \int_0^t e^{\kappa(z)(T-u)}\, dM(z)_u - \int_0^t \xi(z)_u\, dS_u, \qquad (6.11)$$

hence $L(z)$ is a local martingale. Moreover, by the preceding lemma, and noting that $S = M(1)$ as well as $\kappa(1) = 0$ since S is a martingale,

$$\langle S, L(z) \rangle_t = \langle M(1), L(z) \rangle_t$$

$$= \int_0^t e^{\kappa(z)(T-u)}\, d\langle M(1), M(z) \rangle_u - \int_0^t \xi(z)_u\, d\langle M(1), M(1) \rangle_u$$

$$= (\kappa(1+z) - \kappa(z) - \gamma(z)\kappa(2)) \int_0^t e^{\kappa(z)(T-u)} S_u^{z+1}\, du$$

$$= 0.$$

It can be shown that L is square-integrable using estimates based on similar calculations. The last claim follows by combining (6.11) with $\langle L(y), S \rangle = 0$ and Lemma 6.8 since

$$\langle L(y), L(z) \rangle_t = \left\langle L(y), \int_0^{\cdot} e^{\kappa(z)(T-u)}\, dM(z)_u \right\rangle_t$$

$$= \int_0^t e^{\alpha(y,z)(T-u)}\, d\langle M(y), M(z) \rangle_u$$

$$- \int_0^t \gamma(y)\, e^{\alpha(y,z)(T-u)} S_u^{y-1}\, d\langle M(z), M(1) \rangle_u$$

$$= \beta(y,z) \int_0^t e^{\alpha(y,z)(T-u)} S_u^{y+z}\, du. \qquad \square$$

Having found the KW decomposition for power options, the decomposition for general square-integrable claims follows by a stochastic Fubini argument. We only give a sketch of the proof, and refer to Hubalek *et al.* (2006) for full details. Note that they do not assume that S is a martingale, so their estimates in our context are relatively more straightforward but still lengthy.

Proposition 6.10. *KW decomposition for general claims.* *Let $f(S_T)$ be a square-integrable claim which can be represented as in (6.7), and let $H = E[f(S_T)|\mathcal{F}.]$. Then*

$$H = H_0 + \int \xi \, dS + L,$$

where, in the notation of Proposition 6.9,

$$H := \int H(z)\, \Pi(dz),$$

$$\xi := \int \xi(z)\, \Pi(dz),$$

$$L := \int L(z)\, \Pi(dz) = H - H_0 - \int \xi \, dS.$$

Here $\xi \in L^2(S)$, and $L \in \mathcal{M}_0^2$ with $\langle S, L \rangle = 0$. All three processes H, ξ, and L are real-valued.

Proof. (sketch) It can be shown that we may apply Fubini's theorem for stochastic integrals (see Protter (2005), Theorem IV.65) to get

$$\int \int \xi(z)\, dS\, \Pi(dz) = \int \int \xi(z)\, \Pi(dz)\, dS = \int \xi \, dS.$$

By the definition of H, ξ, and L, it then follows from Proposition 6.9 that the KW decomposition of H is indeed as stated. In particular, this decomposition is unique, and as $f(S_T)$ is a real-valued random variable, the three mentioned processes must be real-valued as well. □

After these preparations, we can finally give the proof to our main Theorems 6.5 and 6.6.

Proof. (sketch) Theorem 6.5 follows by orthogonal projection directly from the KW decomposition for H, Proposition 6.10. Regarding

Theorem 6.6, first observe that

$$R\left(\vartheta^H\right) = E\left[L_T^2\right]$$

$$= \int \int E\left[\langle L\left(y\right), L\left(z\right)\rangle_T\right] \Pi\left(dy\right)\Pi\left(dz\right)$$

$$= \int \int R(y,z)\,\Pi\left(dy\right)\Pi\left(dz\right).$$

Hence we get by (6.10) and Fubini's theorem that

$$R(y,z) = E\left[\langle L\left(y\right), L\left(z\right)\rangle_T\right]$$

$$= E\left[\beta\left(y,z\right)\int_0^T e^{\alpha(y,z)(T-t)}S_t^{y+z}\,dt\right]$$

$$= \beta\left(y,z\right)\int_0^T e^{\alpha(y,z)(T-t)}E\left[S_t^{y+z}\right]\,dt$$

$$= S_0^{y+z}\beta\left(y,z\right)\int_0^T e^{\kappa(y+z)t+\alpha(y,z)(T-t)}\,dt.$$

$\square$

The main virtue of the pricing formula (6.9) is that it involves the cumulant generating function $\kappa\left(z\right)$ which often is of a simpler form than the distribution function of X_1. Moreover, the integral in (6.9) can sometimes easily be evaluated by numerical methods involving Fast Fourier Transforms. In some rare examples we can calculate this integral explicitly.

Example 6.11. Black-Scholes option pricing formula. Consider the case where S is a geometric Brownian motion, i.e.

$$X = \sigma B - \frac{1}{2}\sigma^2 t.$$

We have

$$\kappa\left(z\right) = \frac{1}{2}\sigma^2\left(z^2 - z\right),\ \gamma\left(z\right) = z.$$

Of course, in this case the Laplace method does not have an advantage over alternative methods since the density and the characteristic function of X_1 are the same.

Consider a call option with strike one, time to maturity $\tau = T - t$ and current equity price $S_t = s$. We have seen in the previous example that the corresponding measure is $\Pi(dz) = dz/2\pi i(z-1)z$. The option value of this call is, for $R > 1$,

$$V_t = \frac{1}{2\pi i} \int_{R-i\infty}^{R+i\infty} e^{\kappa(z)\tau} s^z \frac{dz}{(z-1)z}$$

$$= \frac{1}{2\pi i} \int_{R-i\infty}^{R+i\infty} e^{\kappa(z)\tau} s^z \frac{dz}{z-1} - \frac{1}{2\pi i} \int_{R-i\infty}^{R+i\infty} e^{\kappa(z)\tau} s^z \frac{dz}{z}$$

$$=: V_t^{(+)} - V_t^{(-)}.$$

The second integral also exists for $R = 0$, is independent of $R \geq 0$, and we have

$$V_t^{(-)}(x) = \frac{1}{2\pi i} \int_{-i\infty}^{+i\infty} e^{\kappa(z)\tau} s^z \frac{dz}{z}$$

$$= \frac{1}{2\pi i} \int_{-i\infty}^{+i\infty} e^{\frac{1}{2}\sigma^2 \tau z^2} e^{xz} \frac{dz}{z}$$

with

$$x = \log s - \frac{1}{2}\sigma^2 \tau.$$

Setting $z = iu$ and differentiating yields

$$\frac{d}{dx} V_t^{(-)}(x) = \frac{1}{2\pi} \int_{-\infty}^{+\infty} e^{-\frac{1}{2}\sigma^2 \tau u^2} e^{iux} du$$

$$= \frac{1}{\sqrt{2\pi}} e^{-\frac{1}{2}\frac{x^2}{\sigma^2 \tau}},$$

hence $V_t^{(-)}(x) = \Phi(\sigma\sqrt{\tau}; x)$, the normal distribution cdf with mean zero and variance $\sigma^2 \tau$.

Similarly, by substituting $z \mapsto z - 1$, and noting that the integral exists in the limit for $R = 1$,

$$V_t^{(+)}(x) = \frac{1}{2\pi i} \int_{-i\infty}^{+i\infty} e^{\kappa(z)\tau} s^z \frac{dz}{z}$$

$$= \frac{s}{2\pi i} \int_{-i\infty}^{+i\infty} e^{\frac{1}{2}\sigma^2 \tau z^2} e^{(x+\sigma^2\tau)z} \frac{dz}{z},$$

hence

$$V_t^{(+)}(x) = s\,\Phi\left(\sigma\sqrt{\tau}; x + \sigma^2\tau\right).$$

For a general strike price $K > 0$ we get, using $C(\zeta; K) = K C(\zeta - \log K; 1)$, the (discounted) Black-Scholes option pricing formula

$$V_t = S_t \, \Phi\left(\sigma\sqrt{\tau}; x_+\right) - K\Phi\left(\sigma\sqrt{\tau}; x_-\right)$$

with $x_+ = \log(S_t/K) + \frac{1}{2}\sigma^2\tau$, $x_- = x_+ - \sigma^2\tau$. Similarly, we get

$$\vartheta_t^H = \Phi\left(\sigma\sqrt{\tau}; x_+ - \frac{1}{2}\sigma^2\tau\right).$$

Since geometric Brownian motion induces a complete market, the variance of hedging error $R\left(\vartheta^H\right)$ is equal to zero.

6.3 Valuation and hedging via integro-differential equations

In this section, let the dynamics of the discounted price process S under a martingale measure P be given as

$$S = S_0 \exp(X)$$

for some Lévy process X with jump measure μ^X and compensator ν^X. Assume that S is a square-integrable martingale, so that X satisfies (EM) for $\beta = 2$. We write the dynamics of S by (4.7) as

$$\frac{dS}{S_-} = \sigma\, dB + d\left(e^x - 1\right) * \left(\mu^X - \nu^X\right).$$

6.3.1 *Feynman-Kac formula for the value function*

Consider a European option $H = h(S_T)$ where the function h satisfies a Lipschitz condition

$$|h(x) - h(y)| \le c\,|x - y|.$$

The value function associated with the option H is defined as

$$V_t = E\left[h(S_T)|\, \mathcal{F}_t\right].$$

If either $\sigma > 0$ or there exists $\beta > 0$ such that

$$\lim_{\varepsilon \searrow 0} \inf \varepsilon^{-\beta} \int_{-\varepsilon}^{\varepsilon} |x|^2 \, \nu(dx) > 0, \tag{6.12}$$

then there exists a $\mathcal{C}^{1,2}$-function $v = v(t, s)$ such that $V_t = v(t, S_t)$. This condition holds if $\nu(dx) \sim c/x^{1+\beta}$ near zero, but e.g. not for the variance gamma model. See Cont & Voltchkova (2005) for a discussion as well as the existence of viscosity solutions for the resulting integro-partial differential equation in case of $\sigma = 0$.

Assuming that condition (6.12) holds enables us to apply Itô's formula to $v(t, S_t)$ to get

$$dv = \frac{\partial v}{\partial t} dt + \frac{\sigma^2}{2} S_t^2 \frac{\partial^2 v}{\partial s^2} dt + \frac{\partial v}{\partial s} dS_t$$
$$+ d\left(v(t, S_{t-} e^x) - v(t, S_{t-}) - S_{t-} (e^x - 1) \frac{\partial v}{\partial s} (t, S_t) \right) * \mu^X.$$

It is worth noting that we can write S_t instead of S_{t-} within dt-terms since as the local martingale S jumps only countably often, the jumps occur on a dt-zero set. Using $\mu^X = \left(\mu^X - \nu^X \right) + \nu^X$, the r.h.s. can be expressed as the sum of a local martingale and a predictable finite variation process A whose differential is

$$dA_t = \left(\frac{\partial v}{\partial t} + \frac{\sigma^2}{2} S_t^2 \frac{\partial^2 v}{\partial s^2} \right) dt$$
$$+ \int \left(v(t, S_t e^x) - v(t, S_t) - S_t (e^x - 1) \frac{\partial v}{\partial s} v(t, S_t) \right) \nu(dx)\, dt.$$

As the value process V is a martingale, the finite variation part A must vanish as the decomposition of a special semi-martingale is unique, resulting in a backward integro-partial differential equation with terminal condition $v(T, s) = h(s)$. This equation can be re-written as a forward equation by considering the time to maturity $\tau = T - t$ instead of real time t, and taking the logarithmic price $x = \log(s/S_0)$ instead of the stock price. Setting $u(\tau, x) = v(T - \tau, S_0 e^x)$ and $f(x) = h(S_0 e^x)$, this substitution yields the equation

$$\frac{\partial u}{\partial \tau} = Lu,$$
$$u(0, x) = f(x)$$

where $Lu = L^{BS} u + L^{\text{jump}} u$ with

$$L^{BS} u = \frac{\sigma^2}{2} \left(\frac{\partial^2}{\partial x^2} u - \frac{\partial}{\partial x} u \right),$$
$$L^{\text{jump}} u = \int \left(u(\tau, x + y) - u(\tau, x) - y \frac{\partial u}{\partial x} (\tau, x) \right) \nu(dy).$$

To sum up, the modified value function u satisfies a Feynman-Kac-type equation involving a non-local term which is an integral with respect to the Lévy measure ν.

6.3.2 Computation of the optimal hedging strategy

We will calculate the optimal mean-variance hedge in an exponential Lévy-model. Recall that the dynamics of S can be written as

$$dS/S_- = \sigma \, dB + (e^x - 1) * d\left(\mu^X - \nu^X\right) := dY,$$

and note that

$$y * \left(\mu^Y - \nu^Y\right) = (e^x - 1) * \left(\mu^X - \nu^X\right).$$

The claim we wish to hedge is a square-integrable European option $H = h(S_T)$ and its associated value function is

$$V_t = v(t, S_t) = E\left[h(S_T)\mid \mathcal{F}_t\right].$$

As in the previous section, under condition (6.12) we may apply the Itô formula to get, since V is a martingale,

$$V_t - V_0 = \int_0^t \frac{\partial v}{\partial s}(u, S_u) S_u \sigma \, dB_u$$

$$+ \left[v(u, S_{u-}(1+y)) - v(u, S_{u-})\right] * \left(\mu^Y - \nu^Y\right)_t.$$

It follows that (we can replace S by S_- within dt-terms)

$$d\langle V, Y\rangle = \left(\frac{\partial v}{\partial s}(u, S_-) S_- \sigma^2 + \int \left[v(u, S_-(1+y)) - v(u, S_-)\right] y\, \nu(dy)\right) dt$$

as well as

$$d\langle Y, Y\rangle = \left(\sigma^2 + \int y^2 \nu(dy)\right) dt.$$

Since $dY = dS/S_-$, the optimal mean-variance hedging strategy is, by $\phi = d\langle V, S\rangle / d\langle S, S\rangle$,

$$\phi_u = \left(\sigma^2 + \int y^2 \nu(dy)\right)^{-1}$$

$$\times \left(\sigma^2 \frac{\partial v}{\partial s}(u, S_-) + \frac{1}{S_-} \int \left[v(u, S_-(1+y)) - v(u, S_-)\right] y\, \nu(dy)\right).$$

6.4   Mean-variance hedging of defaultable assets

In this section we discuss the reduced-form approach to credit or mortality risk. Information about economic factors or overall mortality rates is captured in a filtration $\mathbb{F}$ which we assume to be generated by some Brownian motion B, whereas, the information about whether default (or death) has occurred is contained in a filtration $\mathbb{H}$. The all-encompassing filtration generated by $\mathbb{F}$ and $\mathbb{H}$ is denoted by $\mathbb{G}$. All $\mathbb{G}$-martingales can be written as stochastic integrals with respect to B and the counting process martingale M associated to the one jump process indicating default. Finding the martingale representation of the value process associated to defaultable claims is instrumental in finding the KW decomposition which leads to the optimal mean-variance hedging strategy.

6.4.1   *Intensity-based approach*

The time of default is modelled as a random time $\tau > 0$ living on a probability space $(\Omega, \mathcal{G}, P)$, with $P(\tau > t) > 0$, each $t > 0$. We associate with it the stochastic process H given by

$$H_t = \mathbb{I}_{t \geq \tau},$$

and denote by $\mathbb{H}$ the filtration generated by H. This is the smallest filtration which makes τ into a stopping time. We assume that our probability space supports a Brownian motion B, and denote its augmented natural filtration by $\mathbb{F}$. Let $\mathbb{G} = \mathbb{H} \vee \mathbb{F}$, the filtration generated by both $\mathbb{H}$ and $\mathbb{F}$, and let $\mathcal{G}_T = \mathcal{G}$. Our standing assumption is

> The filtration $\mathbb{F}$ is immersed into the $\mathbb{G}$-filtration.

In particular, B remains a Brownian motion in the larger filtration $\mathbb{G}$. This follows since by immersion, B is a $\mathbb{G}$-martingale, with $[B]_t = t$, and hence is a $\mathbb{G}$-Brownian motion by Lévy's characterisation.

The one jump process H is increasing, so the following result is a consequence of Theorem 2.13.

Theorem 6.12. *Compensator for the one jump process.* *There exists an increasing predictable process* Γ *such that*

$$M = H - \Gamma_{\tau \wedge}. \tag{6.13}$$

is a $\mathbb{G}$-*martingale.*

Throughout this section, we will make the following absolute continuity assumption:

There exists a positive $\mathbb{F}$-predictable process μ such that $\Gamma = \int \mu \, dt$.

We call μ the **intensity** of the random time τ. In the context of credit risk μ is called the hazard rate, whereas, in life insurance μ is often referred to as the force of mortality. Observe that the counting process martingale can be written as

$$M_t = H_t - \int_0^{\tau \wedge t} \mu_s \, ds = H_t - \int_0^t (1 - H_s) \, \mu_s \, ds.$$

Example 6.13. Stochastic Gompertz-Makeham model. The force of mortality is modelled as

$$\mu_t = Y_1(t) + Y_2(t) \, e^{\gamma t}, \qquad \gamma > 0,$$

where the stochastic factors Y_1, Y_2 are Cox-Ingersoll-Ross processes, so in particular are positive. The factor Y_1 models the impact of causes of death unrelated to age, like accidents and infectious diseases. Whereas, the exponential term captures the wearing out effects of ageing.

Definition 6.14. The **survival probability process** G is given by

$$G_t := P\left(\tau > t \mid \mathcal{F}_t\right) = e^{-\Gamma_t} = \exp\left(-\int_0^t \mu_s \, ds\right).$$

We can express conditional expectations with respect to the $\mathbb{G}$-filtration in terms of conditional $\mathbb{F}$-expectations as follows, see Bielecki & Rutkowski (2002), Corollary 5.1.1.

Lemma 6.15. *Let Y be a $\mathcal{G}$-measurable and integrable random variable. Then, for $t \leq s$,*

$$E\left[Y \mathbb{I}_{t < \tau \leq s} \mid \mathcal{G}_t\right] = \mathbb{I}_{\tau > t} \frac{E\left[Y \, \mathbb{I}_{t < \tau \leq s} \mid \mathcal{F}_t\right]}{E\left[\mathbb{I}_{t < \tau} \mid \mathcal{F}_t\right]}$$

$$= (1 - H_t) \, e^{\Gamma_t} E\left[Y \, e^{-\Gamma_T} \mid \mathcal{F}_t\right].$$

The following result is a version of the previous lemma for processes, see Bielecki & Rutkowski (2002), Corollary 5.1.3.

Lemma 6.16. *Let Z be an $\mathbb{F}$-predictable, bounded process. Then, for $t \leq T$,*

$$E\left[Z_\tau \mathbb{I}_{t < \tau \leq T} \mid \mathcal{G}_t\right] = (1 - H_t) \, e^{\Gamma_t} E\left[\int_t^T Z_u \, e^{-\Gamma_u} \mu_u \, du \,\middle|\, \mathcal{F}_t\right].$$

6.4.2 *Martingale representation*

Our goal is to establish a martingale representation result with respect to the filtration $\mathbb{G}$ generated jointly by the Brownian motion B and the counting process martingale M. This is relevant for hedging purposes since every square-integrable claim H can be associated with a square-integrable martingale by taking the conditional expectations $E\left[H\,|\,\mathcal{G}_t\right]$. As a prerequisite, we first study the stochastic exponential of M.

Proposition 6.17. *Stochastic exponential of the one jump process.*
The solution to the stochastic differential equation

$$dL_t = -L_{t-}\,dM_t, \quad L_0 = 1, \tag{6.14}$$

is given by

$$L_t = (1 - H_t)\,e^{\Gamma_t}.$$

Moreover, we have that for any $\mathbb{F}$-martingale N the quadratic co-variation $[L, N]$ vanishes.

Proof. We have by integration by parts that

$$
\begin{aligned}
L_t &= 1 + \int_0^t (1 - H_{s-})\,e^{\Gamma_s}\,d\Gamma_s - \int_0^t e^{\Gamma_s}\,dH_s \\
&= 1 + \int_0^t e^{\Gamma_s}(1 - H_{s-})\,d\Gamma_{s\wedge\tau} - \int_0^t e^{\Gamma_s}(1 - H_{s-})\,dH_s \\
&= 1 - \int_0^t e^{\Gamma_s}(1 - H_{s-})\,d\left(H_s - \Gamma_{s\wedge\tau}\right) \\
&= 1 - \int_0^t L_{s-}\,dM_s,
\end{aligned}
$$

which proves the first statement. For the second part, by Brownian martingale representation we write $N = N_0 + \int \vartheta\,dB$, and therefore

$$[L, N] = -\left[\int L_-\,dM, \int \vartheta\,dB\right] = -\int L_-\vartheta\,d\,[M, B] = 0.$$

We have $[M, B] = 0$ since B is continuous and M of finite variation. $\qquad\square$

First we shall prove a representation result for so-called simple claims which have a certain product structure. The rationale for this structure is that $\mathbb{G}$ is the filtration generated by the two independent filtrations $\mathbb{F}$ and $\mathbb{H}$. Later on general claims will be approximated by a sequence of simple claims.

Definition 6.18. A **simple claim** is a random variable of the form $X = (1 - H_T)Y$ for some square-integrable $\mathcal{F}_T$-measurable random variable Y. Setting $\widetilde{Y} = e^{-\Gamma_T}Y$, we can write

$$X = (1 - H_T)\,e^{\Gamma_T}\widetilde{Y} = L_T\widetilde{Y}.$$

We define an $\mathbb{F}$-martingale U and a predictable process $\widetilde{\xi} \in L^2(B)$ by

$$U_t := E\left[e^{-\Gamma_T}Y\,\middle|\,\mathcal{F}_t\right] = E\left[e^{-\Gamma_T}Y\right] + \int_0^t \widetilde{\xi}_u\,dB_u, \qquad (6.15)$$

where the second equality follows since Brownian motion has the PRP. By the immersion property, U is a $\mathbb{G}$-martingale as well.

For simple claims, an explicit martingale representation can be easily found.

Proposition 6.19. *Simple claims.* *Let* $X = (1 - H_T)Y$ *be a simple claim. Then*

$$X = E\left[e^{-\Gamma_T}Y\right] + \int_0^{\tau \wedge T} \xi_t\,dB_t + \int_0^{\tau \wedge T} \zeta_t\,dM_t$$

where $\xi_t = L_{t-}\widetilde{\xi}_t$ *and* $\zeta_t = -U_t L_{t-}$.

Proof. By integration by parts, we get, since $[L, U] = 0$, as well as (6.14) and (6.15),

$$X = L_T U_T = L_0 U_0 + \int_0^T L_{t-}\,dU_t + \int_0^T U_t\,dL_t$$

$$= E\left[e^{-\Gamma_T}Y\right] + \int_0^{\tau \wedge T} L_{t-}\widetilde{\xi}_t\,dB_t - \int_0^{\tau \wedge T} U_t L_{t-}\,dM_t. \qquad \square$$

A general representation result can now be obtained by approximating $\mathcal{G}_T$-measurable random variables in $L^2(P)$ by simple claims, and using that the spaces of stochastic integrals of square-integrable strategies with respect to B, resp. M, are closed in $L^2(P)$.

Theorem 6.20. *Kusuoka's martingale representation.* *Each square-integrable $\mathbb{G}$-martingale N can be written as*

$$N_t = N_0 + \int_0^t \xi_u\,dB_u + \int_0^t \zeta_u\,dM_u, \qquad (6.16)$$

where $\xi \in L^2(B)$, $\zeta \in L^2(M)$.

The last result is of an abstract nature and does not yield an explicit representation for the strategies ξ and ζ.

6.4.3 *Hedging of insurance claims with longevity bonds*

We will now show how to find the strategies in (6.16) for some generic claims in the context of securitisation of mortality risk. This analysis could easily be interpreted in a credit derivatives framework as well.

A person who is x-years old at time 0 is commonly referred to as (x), and r denotes the constant short rate. Assume that we have sold at time 0 one unit of a **pure endowment** with discounted payoff

$$C^{pe} = e^{-rT}\mathbb{I}_{\tau > T} = e^{-rT}(1 - H_T).$$

One unit of cash will be paid to the policyholder given that (x) is alive at maturity T. Similarly, the discounted payoff of a **term insurance** whereby one unit is paid on the event of death given that this happens before maturity is

$$C^{ti} = e^{-r\tau}\mathbb{I}_{\tau \leq T} = e^{-r\tau}H_T.$$

Finally, we consider a **longevity annuity** with increasing, continuous rate payments equal to $1 - G_t$ while (x) is alive, up to maturity T. Such an instrument rewards longevity relative to the age cohort to which (x) belongs. Its discounted payoff is

$$C^a = \int_0^T e^{-ru}(1 - H_u)(1 - G_u)\,du.$$

We now assume that it is possible to trade on the financial market in an instrument called a **longevity bond** which has discounted payoff

$$B_t = \int_0^t e^{-ru}G_u\,du.$$

The payment generated by this bond has the form of an annuity where the declining rate of payment is given by the survival probability for the age cohort of (x). It does not depend on the individual life history of (x), in contrast to the payouts of the claims considered. The (discounted) value process associated with the longevity bond is given by the conditional expectation

$$V_t = E\left[\int_0^T e^{-ru}G_u\,du\,\middle|\,\mathcal{G}_t\right].$$

We implicitly assume that P is a pricing measure, reflecting the market price of risk. Our goal is to hedge the risk exposure from having sold either a pure endowment, a term insurance or the longevity annuity by trading dynamically in the longevity bond with value process V.

Let us collect some technical notations and assumptions: We assume

$$e^{\Gamma_T} \in L^2(P). \tag{6.17}$$

The space Θ of admissible strategies consists of all predictable ϑ such that

$$E\left[\int_0^T \vartheta_s^2 \, d\,[V]_s\right] < \infty.$$

If $\vartheta \in \Theta$, then $\int \vartheta \, dV$ is a square-integrable martingale. We use the mean-variance hedging approach: for a given claim $C \in L^2(P)$, we want to solve

$$\min_{c,\vartheta} E\left[\left(C - c - \int_0^T \vartheta_s \, dV_s\right)^2\right],$$

where we minimise over all constants c and $\vartheta \in \Theta$. It results by Theorem 6.2 that the fair price is given by $c = E\,[C]$, and the optimal hedging strategy $\vartheta^* \in \Theta$ can be found via the KW decomposition

$$E\,[C|\,\mathcal{G}_t] = c + \int_0^t \vartheta_s^* \, dV_s + V_t^{\perp}, \tag{6.18}$$

where $V^{\perp}$ is a martingale strongly orthogonal to V (i.e. $VV^{\perp}$ is a local martingale).

The decomposition (6.18) of the martingales associated to the various claims can be derived by simple algebra once we have established representations of the martingales $E\,[C|\,\mathcal{G}.]$ and V in terms of stochastic integrals with respect to the Brownian motion B and the counting process martingale M.

Proposition 6.21. *Pure endowment.* *The KW decomposition for a pure endowment C^{pe} is*

$$E\left[e^{-rT}\left(1 - H_T\right)|\,\mathcal{G}_t\right] = e^{-rT}E\left[e^{-\Gamma_T}\right] + \int_0^t \alpha_s^B \, dB_s + \int_0^t \alpha_s^M \, dM_s,$$

where the predictable integrands α^B and α^M are given as

$$\alpha_s^B = \tilde{\xi}_s L_{s-} \tag{6.19}$$
$$\alpha_s^M = -U_s L_{s-}. \tag{6.20}$$

Here $\tilde{\xi}$ corresponds to the integrand in (6.15) for the choice $Y = e^{-rT}$.

Proof. C^{pe} is a simple claim with $Y = e^{-rT}$. The result follows then from Proposition 6.19. $\qquad\square$

Notation. In the following, FV denotes any non-constant term of finite variation.

Now we turn our attention to term insurance. Let us first observe that by martingale representation, there exist a constant c_1 and a $\vartheta \in L^2(B)$ such that

$$E\left[\int_0^T e^{-ru} e^{-\Gamma_u} d\Gamma_u \,\Big|\, \mathcal{F}_t\right] = c_1 + \int_0^t \vartheta_u \, dB_u. \tag{6.21}$$

Proposition 6.22. *Term insurance.* *The KW decomposition for a term insurance C^{ti} is*

$$E\left[e^{-r\tau} H_T \,\big|\, \mathcal{G}_t\right] = c_1 + \int_0^t \beta_s^B \, dB_s + \int_0^t \beta_s^M \, dM_s \,,$$

where

$$\beta_s^B = L_{s-} \vartheta_s, \tag{6.22}$$

$$\beta_u^M = e^{-r(s \wedge \tau)} - L_{s-}\left(c_1 + \int_0^s \vartheta_u \, dB_u - \int_0^s e^{-ru} e^{-\Gamma_u} d\Gamma_u\right). \tag{6.23}$$

Proof. We write

$$E\left[e^{-r\tau} H_T \,\big|\, \mathcal{G}_t\right] = H_t E\left[e^{-r\tau} H_T \,\big|\, \mathcal{G}_t\right] + (1 - H_t) E\left[e^{-r\tau} H_T \,\big|\, \mathcal{G}_t\right]$$

and find the canonical decompositions of the two terms on the r.h.s. into a local martingale and a finite variation part separately. As the conditional expectation on the l.h.s. gives a martingale, the FV-terms have to vanish.

Since $H_t H_T = H_t$, and $H_t e^{-r\tau}$ is $\mathcal{G}_t$-measurable, we get for the first term by integration by parts

$$H_t E\left[e^{-r\tau} H_T \,\big|\, \mathcal{G}_t\right] = H_t e^{-r\tau} = H_t e^{-r(t \wedge \tau)}$$

$$= 0 + \int_0^t H_{s-} \, de^{-r(s \wedge \tau)} + \int_0^t e^{-r(s \wedge \tau)} \, dH_s$$

$$= 0 + \int_0^t e^{-r(s \wedge \tau)} \, dM_s + \text{FV}. \tag{6.24}$$

Here we have used that $H_s = 0$ on $[0, \tau)$, and that $H_s = M_s + \Gamma_{s \wedge \tau}$. As for the second term, we get by Lemma 6.16 that

$$(1 - H_t) E\left[e^{-r\tau} H_T \,\big|\, \mathcal{G}_t\right] = (1 - H_t) E\left[\int_t^T e^{-rs} e^{\Gamma_t - \Gamma_s} d\Gamma_s \,\Big|\, \mathcal{F}_t\right]$$

$$= L_t E\left[\int_t^T e^{-rs} e^{-\Gamma_s} d\Gamma_s \,\Big|\, \mathcal{F}_t\right].$$

Again by integration by parts as well as the martingale representation (6.21),

$$L_t E\left[\int_t^T e^{-rs}e^{-\Gamma_s}\,d\Gamma_s \,\middle|\, \mathcal{F}_t\right]$$

$$= L_t\left(E\left[\int_0^T e^{-rs}e^{-\Gamma_s}\,d\Gamma_s \,\middle|\, \mathcal{F}_t\right] - \int_0^t e^{-rs}e^{-\Gamma_s}\,d\Gamma_s\right)$$

$$= c_1 + \int_0^t \psi_s\,dB_s + \int_0^t \eta_s\,dM_s + \text{FV}\,, \qquad (6.25)$$

where

$$\psi_s = L_{s-}\vartheta_s,$$

$$\eta_s = -L_{s-}\left(c_1 + \int_0^s \vartheta_v\,dB_v - \int_0^s e^{-rv}e^{-\Gamma_v}\,d\Gamma_v\right).$$

The result now follows by combining (6.24) and (6.25), and setting the FV-terms to zero. □

By martingale representation, for each $u \in (t, T]$ there exists a constant c_u and a $\theta_{u,\cdot} \in L^2(B)$ such that

$$E\left[e^{-ru}\left(1 - G_u\right)e^{-\Gamma_u}\,\middle|\,\mathcal{F}_t\right] = c_u + \int_0^t \theta_{u,s}\,dB_s. \qquad (6.26)$$

We set $c_2 = \int_t^T c_u\,du < \infty$. Note that the $\int \theta_{u,\cdot}\,dB$ are bounded martingales, uniformly in u.

Proposition 6.23. *Longevity annuity.* *The KW decomposition of a longevity annuity C^a is*

$$E\left[\int_0^T e^{-ru}\left(1 - H_u\right)\left(1 - G_u\right)du\,\middle|\,\mathcal{G}_t\right] = c_2 + \int_0^t \gamma_s^B\,dB_s + \int_0^t \gamma_s^M\,dM_s\,,$$

where the predictable integrands are given as

$$\gamma_s^B = L_{s-}\int_t^T \theta_{u,s}\,du, \qquad (6.27)$$

$$\gamma_s^M = -L_{s-}\int_t^T\left(c_u + \int_0^s \theta_{u,v}\,dB_v\right)du. \qquad (6.28)$$

Proof. By Lemma 6.15, we have for $u \in (t, T]$

$$E\left[e^{-ru}\left(1 - H_u\right)\left(1 - G_u\right)\middle|\mathcal{G}_t\right] = \left(1 - H_t\right)E\left[e^{-ru}\left(1 - G_u\right)e^{\Gamma_t - \Gamma_u}\middle|\mathcal{F}_t\right]$$

$$= L_t E\left[e^{-ru}\left(1 - G_u\right)e^{-\Gamma_u}\middle|\mathcal{F}_t\right].$$

By integration by parts and (6.26),

$$L_t E\left[e^{-ru}\left(1 - G_u\right)e^{-\Gamma_u}\middle|\mathcal{F}_t\right] = c_u + \int_0^t \phi_{u,s}\,dB_u + \int_0^t \nu_{u,s}\,dM_u, \quad (6.29)$$

where

$$\phi_{u,s} = L_{s-}\,\theta_{u,s},$$

$$\nu_{u,s} = -L_{s-}\left(c_u + \int_0^s \theta_{u,v}\,dB_v\right).$$

By straightforward estimates, it can be shown that we may apply the stochastic Fubini theorem (see Protter (2005), Theorem IV.65) to get from (6.29)

$$\int_0^T \left(1 - H_t\right) E\left[e^{-ru}\left(1 - G_u\right)e^{\Gamma_t - \Gamma_u}\middle|\mathcal{F}_t\right]\,du$$

$$= \int_t^T \left(1 - H_t\right) E\left[e^{-ru}\left(1 - G_u\right)e^{\Gamma_t - \Gamma_u}\middle|\mathcal{F}_t\right]\,du + \mathrm{FV}$$

$$= \int_t^T \left(c_u + \int_0^t \phi_{u,s}\,dB_s + \int_0^t \nu_{u,s}\,dM_s\right)\,du + \mathrm{FV}$$

$$= c_2 + \int_0^t \gamma_s^B\,dB_s + \int_0^t \gamma_s^M\,dM_s + \mathrm{FV},$$

where the predictable integrands γ^B, γ^M are given as

$$\gamma_s^B = \int_t^T \phi_{u,s}\,du, \qquad \gamma_s^M = \int_t^T \nu_{u,s}\,du.$$

This ends the proof. $\qquad\qquad\square$

Finally, we turn to the longevity bond. By martingale representation, for each $u \in (t, T]$ there exists a constant k_u and a predictable $\xi_{u,\cdot}$ such that

$$E\left[e^{-ru}G_u\middle|\mathcal{F}_t\right] = k_u + \int_0^t \xi_{u,s}\,dB_s. \quad (6.30)$$

We set $c_3 = \int_t^T k_u\,du$.

Proposition 6.24. Longevity bond. *The KW decomposition of the longevity bond is*

$$V_t = E\left[\int_0^T e^{-ru}G_u\,du\,\Big|\,\mathcal{G}_t\right] = c_3 + \int_0^t \xi_s\,dB_s,$$

where the predictable integrand ξ is given as

$$\xi_s = \int_t^T \xi_{u,s}\,du.$$

Proof. The discounted survival probability $e^{-ru}G_u$ is bounded and $\mathcal{F}_u$-measurable for every $u \in (t,T]$. Therefore

$$E[e^{-ru}G_u|\mathcal{G}_t] = k_u + \int_0^t \xi_{u,s}\,dB_s.$$

Since G is bounded by one, we have that the $\int \xi_{u,\cdot}\,dB$ are bounded martingales, uniformly in u. We can again apply stochastic Fubini to get

$$E\left[\int_0^T e^{-ru}G_u\,du\,\Big|\,\mathcal{G}_t\right] = \int_t^T E[e^{-ru}G_u|\mathcal{G}_t]du + \mathrm{FV}$$

$$= c_3 + \int_0^t \xi_s\,dB_s + \mathrm{FV}$$

where the predictable integrand ξ is given as

$$\xi_s = \int_t^T \xi_{u,s}\,du.$$

$\square$

Summing up, the various claim payoffs allow for a representation

$$C = c + \int_0^T \delta_u^B\,dB_u + \int_0^T \delta_u^M\,dM_u\,, \tag{6.31}$$

where the integrands δ^B, δ^M are claim-specific and have been obtained in the foregoing propositions. Moreover, the value process of the longevity bond, which serves as hedging instrument, has representation

$$V_t = V_0 + \int_0^t \xi_u\,dB_u. \tag{6.32}$$

As we have seen, the integrands δ^B, δ^M, ξ can all be computed and are therefore be considered to be known quantities. Our goal is now to find the KW decomposition

$$E[C|\mathcal{G}_t] = c + \int_0^t \vartheta_u^*\,dV_u + V_t^\perp, \tag{6.33}$$

where $V^{\perp}$ is a square-integrable martingale, strongly orthogonal to V with decomposition

$$V_t^{\perp} = \int_0^t \delta_u^M \, dM_u.$$

The term $1 + \int_0^t \vartheta_u^* \, dV_u$ can be interpreted as the part of the risk that can be perfectly replicated by means of our optimal hedging strategy ϑ^*, and $V_t^{\perp}$ as the part of the risk that is totally unhedgeable.

The integrand ϑ^* in the KW decomposition (6.33) is determined uniquely by the equation

$$\vartheta^* \xi = \delta^B .$$

Uniqueness should be understood modulo the following equivalence relation: if $\vartheta, \psi \in \Theta$, then

$$\vartheta \sim \psi \quad \text{if} \quad \int_0^T (\vartheta_t - \psi_t)^2 \, d\,[V]_t = 0. \tag{6.34}$$

In particular, the predictable process $\vartheta^* \in \Theta$ is the unique mean-variance hedging strategy for the claim when trading in the underlying longevity bond.

6.5 Quadratic risk-minimisation for payment streams

Mean-variance hedging of payment streams can be undertaken within the setting introduced in Section 6.1. In particular, we assume that $S \in \mathcal{M}_{loc}^2$. The objective is to describe the value process associated with a strategy which perfectly replicates a payment stream. However, a more general definition of a strategy is required as we cannot insist that strategies are self-financing. In the following, alongside trading in the stock S according to ϑ, η units are held in a bank account with $R \equiv 1$.

Definition 6.25. A **strategy** is a process $\varphi = (\vartheta, \eta)$ where $\vartheta \in L^2(S)$ and η is $\mathbb{F}$-adapted such that the **value process**

$$V(\varphi) := \vartheta \, S + \eta$$

is cadlag, and $V_t(\varphi) \in L^2(P)$ for all $t \in [0, T]$. A **payment stream** A is an $\mathbb{F}$-adapted, cadlag process such that $A_t \in L^2(P)$ for all $t \in [0, T]$. The **cost process** associated with φ and A is

$$C(\varphi) = V(\varphi) - \int \vartheta \, dS + A,$$

and the corresponding **risk process** $R(\varphi)$ is given by

$$R_t(\varphi) = E\left[\left(C_T(\varphi) - C_t(\varphi)\right)^2 \middle| \mathcal{F}_t\right].$$

Among all strategies leading to the same terminal value, we are looking for one which has smallest risk process at all points in time prior to maturity.

Definition 6.26. A strategy φ is **risk-minimising** if for all strategies $\widetilde{\varphi}$ with $V_T(\varphi) = V_T(\widetilde{\varphi})$ we have

$$R_t(\varphi) \le R_t(\widetilde{\varphi}) \quad \text{for every } t \in [0,T].$$

We call φ **mean self-financing** if the cost process $C(\varphi)$ is a martingale.

Lemma 6.27. *If the strategy φ is risk-minimising, then φ is mean self-financing.*

Proof. Let $s \in [0,T]$ be arbitrary, and define another strategy $\widetilde{\varphi}$ by setting $\widetilde{\vartheta} = \vartheta$, and taking $\widetilde{\eta}$ such that $V_t(\widetilde{\varphi}) = V_t(\varphi)$ for $t \in [0,s)$, and, for $t \in [s,T]$,

$$V_t(\widetilde{\varphi}) = E\left[V_T(\varphi) - \int_t^T \vartheta_u\, dS_u + A_T - A_t \middle| \mathcal{F}_t\right].$$

Then $\widetilde{\varphi}$ is a strategy with $V_T(\varphi) = V_T(\widetilde{\varphi})$ and $C_T(\varphi) = C_T(\widetilde{\varphi})$. Furthermore, we have $C_s(\widetilde{\varphi}) = E\left[C_T(\widetilde{\varphi})\middle| \mathcal{F}_s\right]$ by construction of $\widetilde{\varphi}$. Therefore,

$$C_T(\varphi) - C_s(\varphi) = C_T(\widetilde{\varphi}) - C_s(\widetilde{\varphi}) + E\left[C_T(\widetilde{\varphi})\middle| \mathcal{F}_s\right] - C_s(\varphi),$$

yielding

$$R_s(\varphi) = R_s(\widetilde{\varphi}) + \left(E\left[C_T(\varphi)\middle| \mathcal{F}_s\right] - C_s(\varphi)\right)^2.$$

Since φ is risk-minimising, and hence $R_s(\varphi) - R_s(\widetilde{\varphi}) \le 0$, we conclude that

$$E\left[C_T(\varphi)\middle| \mathcal{F}_s\right] = C_s(\varphi). \qquad \square$$

Consider the square-integrable martingale V^A, given by

$$V_t^A = E\left[A_T\middle| \mathcal{F}_t\right],$$

which has a KW decomposition

$$V^A = V_0^A + \int \vartheta^A\, dS + L \tag{6.35}$$

with $\vartheta^A \in L^2(S)$ and $L \in \mathcal{M}_0^2$, strongly orthogonal to each element of the stable subspace $\mathcal{S}(S)$.

Theorem 6.28. *Risk-minimisation for payment streams.* *Every payment stream A admits a unique risk-minimising strategy φ^* with $V_T(\varphi^*) = 0$. In terms of the decomposition (6.35), φ^* is given as*

$$\varphi^* = (\vartheta^*, \eta^*) = \left(\vartheta^A, V^A - A - \vartheta^A S\right).$$

The associated risk process is

$$R_t(\varphi) = E\left[\left(L_T - L_t\right)^2 \middle| \mathcal{F}_t\right].$$

Proof. **Existence:** By (6.35), we get

$$A_T = V_t^A + \int_t^T \vartheta_u^A \, dS_u + L_T - L_t.$$

For any strategy φ with $V_T(\varphi) = 0$, we therefore have

$$C_T(\varphi) - C_t(\varphi) = -V_t(\varphi) - \int_t^T \vartheta_u \, dS_u + A_T - A_t$$

$$= \left(V_t^A - A_t - V_t(\varphi)\right) + \int_t^T \left(\vartheta_u^A - \vartheta_u\right) dS_u + \left(L_T - L_t\right).$$

Since L is strongly orthogonal to each element of $\mathcal{S}(S)$, and the first term on the r.h.s. is $\mathcal{F}_t$-measurable, it results that

$$R_t(\varphi) = \left(V_t^A - A_t - V_t(\varphi)\right)^2 + E\left[\int_t^T \left(\vartheta_u^A - \vartheta_u\right)^2 d[S]_u \middle| \mathcal{F}_t\right]$$

$$+ E\left[\left(L_T - L_t\right)^2 \middle| \mathcal{F}_t\right].$$

Note that the last term is independent of the strategy φ. The first two terms equal zero if we first choose $\vartheta = \vartheta^A$, and then η such that $V(\varphi) = V^A - A$.

Uniqueness: A risk-minimising strategy $\varphi = (\vartheta, \eta)$ with $V_T(\varphi) = 0$ minimises in particular $R_0(\cdot)$, and hence $\vartheta = \vartheta^A$. Moreover, $C(\varphi)$ is a martingale by Lemma 6.27, which implies that $V_t(\varphi) = V^A - A$, and therefore $\eta = V^A - A - \vartheta^A S$. $\quad\square$

6.6 Notes and further reading

A survey of various quadratic approaches to hedging (not only under the risk-neutral measure) can be found in Schweizer (2001). The Laplace method for option pricing has been pioneered by Raible (2000), whereas mean-variance hedging in this context has been studied by Hubalek *et al.*

(2006). The integro-differential approach comes from Cont & Tankov (2003) as well as Cont & Voltchkova (2005), and is the subject of the monograph Boyarchenko & Levendorskiĭ (2002). Mean-variance hedging with options in a jump-diffusion framework has been considered in Cont *et al.* (2007), and mean-variance hedging for affine stochastic volatility models by Kallsen & Pauwels (2010a), (2010b). For monographs about credit risk modelling which have material relevant for hedging see Bielecki & Rutkowski (2002) as well as Jeanblanc *et al.* (2009). Risk-minimising hedging strategies for unit-linked life insurance contracts are designed in Møller (1998). A local risk-minimising approach to hedging of defaultable assets can be found in Biagini & Cretarola (2009), and mean-variance hedging of insurance claims in Dahl *et al.* (2008) as well as Biagini *et al.* (2011). Hedging of payment streams is particularly important in insurance applications, and has been studied by Møller (2001).

Chapter 7

Entropic Valuation and Hedging

In this chapter we approach hedging in incomplete markets by formulating a utility indifference criterion under the statistical measure P, employing the exponential utility function. To study qualitative properties of the indifference price and optimal hedge, we pursue a duality approach which naturally leads us to a discussion of the minimal entropy martingale measure. When the agent's risk aversion parameter tends to zero, we asymptotically recover the risk-neutral mean-variance hedging approach discussed in Chapter 6, formulated under the entropy measure.

7.1 Exponential utility indiffence pricing

Consider an agent who sells a claim with stochastic payoff B for a deterministic premium π. In the absence of dynamic trading opportunities, the utility indifference price π is defined as the solution to the equation

$$U(c) = E[U(c + \pi - B)]$$

with a utility function U and a constant wealth level of c. In actuarial mathematics, such a valuation method is known as the 'premium principle of equivalent utility'. In this chapter, we will focus on the **exponential utility**,

$$U_\alpha(x) = 1 - \exp(-\alpha x) \qquad \text{for } \alpha > 0. \tag{7.1}$$

The choice of the exponential utility function often has some computational advantages; yet it is sometimes questioned due to the following observation. Consider the claim B: with probability $1/2$ the agent receives an infinite amount of money; whereas with probability $1/2$ she has to make a payment of a units. If $a > 1/\alpha \log 2$, then an agent with exponential utility function would reject entering this bet no matter how high her wealth level c is, since then

$$U_\alpha(c) > E[U_\alpha(c - B)]$$

for all $c > 0$. However, this appears to be counter-intuitive since for a very rich person, where a is just a tiny fraction of her total wealth, we would anticipate this bet would look quite attractive. However, since $1/\alpha \log 2 \to \infty$ as $\alpha \downarrow 0$, the effect becomes ever less pronounced for small α. This observation provides a motivation for why we will later focus on an asymptotic exponential utility indifference price for $\alpha \downarrow 0$.

In the presence of a financial market, investors can reduce their risk exposure to the terminal liability B by dynamically trading in a risky asset with discounted price process S. The trading strategy ϑ must be chosen from a space Θ of admissible strategies which shall be specified below. Let C be any claim and consider the exponential utility function (7.1). We set

$$u(C; \alpha) = \sup_{\vartheta \in \Theta} E\left[U_\alpha\left(C + \int_0^T \vartheta_t \, dS_t\right)\right].$$

In the next definition, we compare this expression for the claims $C_1 = c$ and $C_2 = c + \pi - B$.

Definition 7.1. Let B be any claim, c an initial wealth level. If there is a unique solution $\pi = \pi(B; \alpha)$ to the equation

$$u(c; \alpha) = u(c + \pi - B; \alpha) \tag{7.2}$$

we call it the **utility indifference price** for B.

The idea is that the investor is indifferent about just investing in S or selling in addition the claim B for the premium π. Let us, for the sake of clarity, restate equation (7.2) after a little algebra as

$$\sup_{\vartheta \in \Theta} E\left[-\exp\left(-\alpha\left(\int_0^T \vartheta_t \, dS_t\right)\right)\right]$$

$$= \sup_{\vartheta \in \Theta} E\left[-\exp\left(-\alpha\left(\pi + \int_0^T \vartheta_t \, dS_t - B\right)\right)\right].$$

In particular, due to the multiplicative property of the exponential function, π does not depend on the wealth level c.

7.2   The minimal entropy martingale measure

It is difficult to obtain properties of the indifference price just by using its definition. In this chapter, we shall proceed via a duality result which shall be established later on. First, we relate maximising the expected exponential utility from pure investment to minimising an entropy functional over all equivalent martingale measures for S. The general result regarding investment plus issuing a claim then follows by a change of measure.

Definition 7.2. The **relative entropy** $H(Q, P)$ of a probability measure Q with respect to a probability measure P is given as

$$H(Q,P) = \begin{cases} E_P\left[\dfrac{dQ}{dP}\log\dfrac{dQ}{dP}\right] & \text{if } Q \ll P \\ +\infty & \text{otherwise} \end{cases}.$$

While the concept of relative entropy originates from information theory, we will have no need of this interpretation here. The main reason why relative entropy relates to the exponential utility function is by convex duality.

In the following, P, Q, and R all denote probability measures. For the proof of the following elementary properties, the reader is referred to Ihara (1992).

Lemma 7.3. *The relative entropy $H(Q, P)$ has the following properties:*

(i) $H(Q, P) \geq 0$;
(ii) $H(Q, P) = 0$ if and only if $Q = P$;
(iii) *The mapping given by $Q \to H(Q, P)$ is strictly convex.*

Definition 7.4. The **minimal entropy martingale measure** Q^0 (in short: entropy measure) attains

$$H(Q^0, P) = \min_{Q \in \mathcal{M}} H(Q, P).$$

We impose the following standing assumption:

S is a locally bounded semi-martingale.

Recall that $Q \in \mathcal{M}$ if $Q \ll P$ and S is a local Q-martingale, so $\mathcal{M}$ denotes the space of all absolutely continuous martingale measures for S. Let us give an alternative definition for $\mathcal{M}$ which will be useful in the sequel.

Definition 7.5. Let $0 \le T_1 \le T_2 \le T$ be stopping times such that the stopped process S^{T_2} is bounded and h is a bounded $\mathcal{F}_{T_1}$-measurable random variable. Denote by $\mathcal{V}$ the linear subspace of $L^\infty(\Omega, \mathcal{F}, P)$ spanned by the elementary stochastic integrals of the form $f = h(S_{T_2} - S_{T_1})$. A **martingale measure** for $\mathcal{V}$ is a probability measure $Q \ll P$ with $E_Q[f] = 0$ for all $f \in \mathcal{V}$.

Under our assumption that S is locally bounded, this definition coincides with the previous definition of $\mathcal{M}$ since Q is a martingale measure for $\mathcal{V}$ if and only if S is a local Q-martingale. In the following, we shall write

$$\mathcal{M}^f = \{Q \in \mathcal{M} \,|\, H(Q, P) < \infty\}.$$

Definition 7.6. If both $P \ll R$, $Q \ll R$, then the **total variation distance** between P and Q is defined to be

$$|Q - P| = \int \left| \frac{dQ}{dR} - \frac{dP}{dR} \right| \, dR.$$

While the following inequality is elementary, its proof is nevertheless not easy, see Ihara (1992).

Proposition 7.7. *Variation — entropy inequality. We have*

$$|Q - P|^2 \le 2H(Q, P).$$

Lemma 7.8. *The set $\mathcal{M}$ is convex and closed in variation.*

Proof. This follows from the alternative characterisation of $\mathcal{M}$. Indeed, convexity is clear, and as for closedness, let $(Q_n) \subset \mathcal{M}$ such that $Q_n \to Q$ in total variation. Then the densities converge in $L^1(P)$, and we have for any $f \in \mathcal{V}$ that

$$E_Q[f] = E\left[\frac{dQ}{dP} f \right] = \lim_{n \to \infty} E\left[\frac{dQ_n}{dP} f \right] = 0. \qquad \square$$

Lemma 7.9. *If P, Q, and $R \in \mathcal{M}$ (or are from any convex set of probability measures), we have the **parallelogram identity***

$$H(P, R) + H(Q, R)$$
$$= 2H\left(\frac{P+Q}{2}, R \right) + H\left(P, \frac{P+Q}{2} \right) + H\left(Q, \frac{P+Q}{2} \right).$$

Proof. Convexity ensures that $(P + Q)/2 \in \mathcal{M}$, and the formula follows by writing all terms as expectations with respect to R. $\qquad \square$

Theorem 7.10. Existence and uniqueness of the entropy measure.
If $\mathcal{M}^f \neq \varnothing$ then the minimal entropy martingale measure Q^0 exists and is unique.

Proof. **Uniqueness:** Let $Q_1 \neq Q_2$ be two different minimal entropy martingale measures. Define a probability measure $\widehat{Q}$ by

$$\widehat{Q} = \frac{1}{2}(Q_1 + Q_2).$$

It follows from the strict convexity of the relative entropy that

$$\min_{Q \in \mathcal{M}} H(Q,P) \leq H\left(\widehat{Q},P\right) < \frac{1}{2}\left(H(Q_1,P) + H(Q_2,P)\right)$$

$$= \min_{Q \in \mathcal{M}} H(Q,P)$$

and we arrive at a contradiction.

Existence: Consider an entropy-minimising sequence $(Q_n) \subset \mathcal{M}^f$,

$$H(Q_n,P) \to \inf_{Q \in \mathcal{M}} H(Q,P).$$

By the parallelogram identity, we have

$$H(Q_m,P) + H(Q_n,P) \tag{7.3}$$

$$= 2H\left(\frac{Q_m+Q_n}{2},P\right) + H\left(Q_m,\frac{Q_m+Q_n}{2}\right) + H\left(Q_n,\frac{Q_m+Q_n}{2}\right).$$

Using the convexity of the relative entropy, we get

$$H\left(\frac{Q_m+Q_n}{2},P\right) \to \inf_{Q \in \mathcal{M}} H(Q,P),$$

and therefore the last two terms in (7.3) must converge to zero as $m,n \to \infty$. By the variation–entropy inequality, we conclude that

$$|Q_m - Q_n| \leq \left|Q_m - \frac{Q_m+Q_n}{2}\right| + \left|Q_n - \frac{Q_m+Q_n}{2}\right|$$

$$\leq \sqrt{2H\left(Q_m,\frac{Q_m+Q_n}{2}\right)} + \sqrt{2H\left(Q_n,\frac{Q_m+Q_n}{2}\right)}$$

$$\to 0 \qquad \text{for } n,m \to \infty.$$

As $\mathcal{M}$ is closed in variation, the Q_n's converge in variation to some $Q \in \mathcal{M}$. By Fatou's lemma,

$$H(Q,P) \leq \liminf_{n\to\infty} H(Q_n,P) < \infty.$$

As (Q_n) is an entropy minimising sequence, Q must be the minimal entropy martingale measure. $\qquad\square$

Let us now state some further elementary properties of the relative entropy, for the proofs see Cziszár (1975).

Lemma 7.11. *We have:*

(i) *Let* $\overline{Q} \in \mathcal{M}^f$. *Then* $\overline{Q} = Q^0$ *if and only if*

$$H(Q, P) \geq H(Q, \overline{Q}) + H(\overline{Q}, P) \qquad \text{for all } Q \in \mathcal{M}^f. \qquad (7.4)$$

(ii) *Let* $\mathcal{M}'$ *be a convex subset of* $\mathcal{M}$. *If for all* $Q_1 \in \mathcal{M}'$ *there exists* $\alpha \in (0,1)$ *and* $Q_2 \in \mathcal{M}'$ *such that* $Q^0 = \alpha Q_1 + (1-\alpha) Q_2$, *i.e.* Q^0 *is an* **algebraic inner point** *of* $\mathcal{M}'$, *then* $\mathcal{M}' \subset \mathcal{M}^f$ *and (7.4) holds with equality for all* $Q \in \mathcal{M}'$.

When equality holds in (7.4), the relative entropy can be seen as analogous to the square of the Euclidean norm, and (7.4) corresponds then to Pythagoras' theorem. However, to show that equality holds indeed is often the tricky part of a proof.

Proposition 7.12. *Equivalence of the entropy measure.* *Assume* $\mathcal{M}^f \cap \mathcal{M}^e \neq \varnothing$. *Then* Q^0 *is equivalent to* P.

Proof. As $H(Q^0, P) < \infty$, by definition of relative entropy we have $Q^0 \ll P$. On the other hand, for some $Q \in \mathcal{M}^f \cap \mathcal{M}^e$ we have by Lemma 7.11 (i) that

$$H(Q, Q^0) \leq H(Q, P) - H(Q^0, P).$$

As the r.h.s. of this inequality is finite, the definition of relative entropy implies that $Q \ll Q^0$. Since $Q \sim P$ we get that also $P \ll Q^0$ and therefore $P \sim Q^0$. $\qquad \square$

The following standing assumption will be imposed from now on:

$$\boxed{\mathcal{M}^f \cap \mathcal{M}^e \neq \varnothing.}$$

Under this assumption, we have seen that Q^0 exists uniquely, and is equivalent to P. We now turn to an important structural result which states that the logarithm of the density of the entropy measure for S can always be written as sum of a constant and a stochastic integral with respect to S.

Proposition 7.13. *Csiszár's result. We can write*

$$\frac{dQ^0}{dP} = \exp\left(c + g\right) \qquad Q^0-\text{a.s. (hence also } P-\text{a.s.),}$$

with a constant c and $g \in \overline{V}$ where the closure is taken in $L^1\left(Q^0\right)$.

Proof. We have to show that

$$\log \frac{dQ^0}{dP} \in \overline{\text{span}\left(1 + V\right)}^{L^1\left(Q^0\right)}.$$

Take any $h \in L^\infty\left(Q^0\right)$ (the dual space of $L^1\left(Q^0\right)$) which vanishes on span$\left(1 + V\right)$, i.e.

$$E_{Q^0}\left[h\right] = 0, \quad E_{Q^0}\left[hf\right] = 0 \qquad \text{for all } f \in V. \tag{7.5}$$

By Hahn-Banach (see e.g. Conway (1990)), it suffices to show that we have

$$E_{Q^0}\left[\log \frac{dQ^0}{dP} h\right] = 0. \tag{7.6}$$

Indeed, for any h satisfying (7.5) and $|h| \leq 1$ we define a measure Q via

$$\frac{dQ}{dQ^0} = 1 + h. \tag{7.7}$$

Q is a probability measure since by (7.5),

$$E_P\left[\frac{dQ}{dP}\right] = E_P\left[\frac{dQ^0}{dP}\frac{dQ}{dQ^0}\right] = E_{Q^0}\left[1 + h\right] = 1.$$

It follows from $Q^0 \in \mathcal{M}$ and (7.5) that for all $f \in V$ we have

$$E_Q\left[f\right] = E_{Q^0}\left[\frac{dQ}{dQ^0}f\right] = E_{Q^0}\left[\left(1 + h\right)f\right] = 0.$$

Together with $|h| \leq 1$ we therefore get

$$Q \in \mathcal{M}' = \left\{Q \in \mathcal{M} \,\middle|\, \frac{dQ}{dQ^0} \leq 2\right\}.$$

The measure Q^0 is an algebraic inner point of $\mathcal{M}'$ (in general, it need not be an algebraic inner point of $\mathcal{M}$): for any $Q_1 \in \mathcal{M}'$ there is a $Q_2 \in \mathcal{M}'$ with $dQ_2/dQ^0 := 2 - dQ_1/dQ^0$ and hence $Q^0 = \left(Q_1 + Q_2\right)/2$. Indeed, we have

$$E_P\left[\frac{dQ_2}{dP}\right] = E_{Q^0}\left[\frac{dQ_2}{dQ^0}\right] = 2 - E_{Q^0}\left[\frac{dQ_1}{dQ^0}\right]$$

$$= 2 - E_P\left[\frac{dQ_1}{dP}\right] = 1,$$

and, for any $f \in \mathcal{V}$,

$$E_{Q_2}[f] = E_{Q^0}\left[\frac{dQ_2}{dQ^0}f\right] = 2E_{Q^0}[f] - E_{Q_1}[f] = 0.$$

This implies by Lemma 7.11 (*ii*) that for Q as in (7.7),

$$H(Q,P) = H(Q,Q^0) + H(Q^0,P),$$

or equivalently,

$$E_Q\left[\log\frac{dQ^0}{dP}\right] = E_{Q^0}\left[\log\frac{dQ^0}{dP}\right],$$

or even

$$E_{Q^0}\left[\log\frac{dQ^0}{dP}\left(\frac{dQ}{dQ^0}-1\right)\right] = 0.$$

Equation (7.6) follows now from (7.7). $\square$

By L^1-martingale representation as in Theorem 2.34, we can identify the $L^1(Q^0)$-closure of $\mathcal{V}$ with the terminal values of the stochastic integral processes $\int \vartheta\, dS$ which are Q^0-martingales. Therefore Proposition 7.13 implies:

Theorem 7.14. Density of the entropy measure. *The density of Q^0 can be written in the following form:*

$$\frac{dQ^0}{dP} = \exp\left(c_0 + \int_0^T \vartheta_t^0\, dS_t\right) \tag{7.8}$$

where c_0 is a constant and ϑ^0 a predictable process such that $\int \vartheta^0\, dS$ is a Q^0-martingale.

The converse of this result is not true. There exists a martingale measure whose density can be written as in (7.8) but is not the entropy minimiser, see Schachermayer (2003) for details. A partial converse is given by the following proposition.

Proposition 7.15. Characterisation of entropy measure. *Assume there exists $Q^* \in \mathcal{M}^f \cap \mathcal{M}^e$. Then $Q^* = Q^0$ if and only if*

(i) $dQ^*/dP = \exp\left(c + \int_0^T \eta_t\, dS_t\right)$ *for some constant c and an S-integrable η, and*

(ii) $E_Q\left[\int_0^T \eta_t\, dS_t\right] = 0$ *for $Q = Q^*, Q^0$.*

Proof.   First note that due to our standing assumption and Theorem 7.10 Q^0 exists and is unique, and that necessity follows from Theorem 7.14. It only remains to prove sufficiency. We have by (i) and (ii)

$$H\left(Q^*, P\right) = E_{Q^*}\left[\log \frac{dQ^*}{dP}\right] = c + E_{Q^*}\left[\int_0^T \eta_t\, dS_t\right] = c.$$

On the other hand, we get by the positivity of the relative entropy as well as (i), (ii)

$$H\left(Q^0, P\right) = H\left(Q^0, Q^*\right) + E_{Q^0}\left[\log \frac{dQ^*}{dP}\right]$$

$$\geq c + E_{Q^0}\left[\int_0^T \eta_t\, dS_t\right]$$

$$= c.$$

It follows that

$$H\left(Q^0, P\right) \geq H\left(Q^*, P\right),$$

and we conclude by the uniqueness of Q^0 that $Q^* = Q^0$. $\square$

It is not possible to directly verify condition (ii) in the previous proposition because it involves Q^0 which is a priori not known. To state a version of Proposition 7.15 which can be verified directly, we need an additional integrability condition on the quadratic variation of the stochastic integral process $\int \eta\, dS$. As preparation we state the following useful inequality.

Lemma 7.16. *Let Q be a probability measure and B a random variable that is uniformly bounded from below or in $L^1(Q)$. Then*

$$E_Q\left[B\right] \leq H\left(Q, P\right) - 1 + E_P\left[e^B\right].$$

Proof.   The assumption on B guarantees that the l.h.s. is well defined in $(-\infty, \infty]$. Moreover, we can assume that $Q \ll P$. The convex conjugate of $\exp(x)$ equals $x \log x - x$ for $x > 0$, so Fenchel's inequality yields

$$xy \leq x \log x - x + e^y \qquad \text{for } x > 0,\ y \geq 0.$$

The result follows by choosing $x = dQ/dP$, $y = B$ and taking expectations under P. $\square$

In the following the Orlicz space for the Young function $\exp(\cdot)$ is denoted by

$$L_{\exp} = \{ f \mid E\left[\exp\left(\varepsilon \,|f|\right)\right] < \infty \text{ for some } \varepsilon > 0 \} .$$

Proposition 7.17. *Verification result.* *If an equivalent martingale measure Q^* with finite relative entropy is of the form*

$$\frac{dQ^*}{dP} = \exp\left(c + \int_0^T \eta_t \, dS_t \right)$$

with $\int_0^T \eta_t^2 \, d\,[S]_t \in L_{\exp}(P)$, then $\int \eta \, dS$ is a true Q-martingale for all $Q \in \mathcal{M}^e$ with finite relative entropy, and therefore Q^* is the minimal entropy martingale measure.

Proof. Fix an arbitrary $Q \in \mathcal{M}^f \cap \mathcal{M}^e$. For α small enough, we have by assumption

$$\alpha E_Q \left[\int_0^T \eta_t^2 \, d\,[S]_t \right] \le H\left(Q, P\right) + E\left[\exp\left(\alpha \int_0^T \eta_t^2 \, d\,[S]_t \right) \right] < \infty,$$

where we have used Lemma 7.16. $\int \eta \, dS$ is a true Q-martingale since its square bracket process is integrable, see Theorem 2.15. The last statement follows from Proposition 7.15. □

7.3   Duality results

We now turn our attention to duality results which link exponential utility maximisation with minimisation of an entropy functional over martingale measures. This duality is fundamental to understanding the qualitative properties of the exponential utility indifference price. The representation of the minimal entropy measure in Theorem 7.14 will turn out to be instrumental in the proof of these duality results.

Definition 7.18. We define the following spaces of admissible integrands:

$$\widetilde{\Theta} = \left\{ \vartheta \in L(S) \;\middle|\; \int \vartheta \, dS \text{ is a } Q^0\text{-martingale} \right\},$$

$$\Theta = \left\{ \vartheta \in L(S) \;\middle|\; \int \vartheta \, dS \text{ is a } Q\text{-martingale for all } Q \in \mathcal{M}^f \right\}.$$

More precisely, by a slight abuse of notation, we denote by Θ (similarly for $\widetilde{\Theta}$) the quotient space $\Theta/\sim$ where $\sim$ denotes the equivalence relation

$$\vartheta \sim \psi \quad \text{if} \quad \int_0^T (\vartheta_t - \psi_t)^2 \, d\,[S]_t = 0.$$

Proposition 7.19. *Entropic duality (preliminary version).* *We have*

$$\inf_{\vartheta \in \widetilde{\Theta}} E\left[\exp\left(-\alpha \int_0^T \vartheta_t \, dS_t\right)\right] = \exp\left(-H\left(Q^0, P\right)\right) \qquad (7.9)$$

and the inf *is attained by* $-\frac{1}{\alpha}\vartheta^0 \in \widetilde{\Theta}$ *where* ϑ^0 *is determined by (7.8).*

Proof. Since $H\left(Q^0, P\right) = c_0$ by Theorem 7.14, the r.h.s. of (7.9) equals $\exp(-c_0)$. On the other hand, for any $\vartheta \in \widetilde{\Theta}$ we get by (7.8) and Jensen's inequality

$$E\left[\exp\left(-\alpha \int_0^T \vartheta_t \, dS_t\right)\right]$$

$$= e^{-c_0} E_{Q^0}\left[\exp\left(-\int_0^T \vartheta_t^0 \, dS_t - \alpha \int_0^T \vartheta_t \, dS_t\right)\right]$$

$$\geq e^{-c_0} \exp\left(E_{Q^0}\left[-\int_0^T \vartheta_t^0 \, dS_t - \alpha \int_0^T \vartheta_t \, dS_t\right]\right)$$

$$= e^{-c_0}.$$

The last equality follows because $\int \vartheta^0 \, dS$ is a Q^0-martingale by Theorem 7.14 and $\int \vartheta \, dS$ is a Q^0-martingale since $\vartheta \in \widetilde{\Theta}$. Moreover, the l.h.s. equals $\exp(-c_0)$ when we choose $\vartheta = -\frac{1}{\alpha}\vartheta^0 \in \widetilde{\Theta}$. $\qquad\square$

While this duality result has a very simple proof, it has the conceptual drawback that it seems to anticipate the duality as $\widetilde{\Theta}$ is defined in terms of the minimiser Q^0 of the dual optimisation problem $\inf_{Q \in \mathcal{M}^f} H\left(Q, P\right)$. Another problem with $\widetilde{\Theta}$ is that it is not well suited to hedging as we will need to consider later on a whole family $\{Q^\alpha\}$ of perturbed entropy measures and not only one single element Q^0. However, it turns out that the entropic duality is robust with respect to the choice of admissible spaces of strategies and in particular holds for Θ as well. First we need a preliminary result, and recall that for $t \in [0, T]$, $\mathcal{T}_t$ denotes the set of all stopping times

τ with values in the interval $[t, T]$. Moreover, for $Q \in \mathcal{M}$ with density process Z we let

$$\log \widetilde{Z}_t := E_Q \left[\log \frac{Z_T}{Z_t} \,\Big|\, \mathcal{F}_t \right].$$

We denote the corresponding quantities for Q^0 by Z^0 and $\widetilde{Z}^0$.

Lemma 7.20. *For every $Q \in \mathcal{M}^e$, the family $\big\{ \log \widetilde{Z}^0_\tau \big\}_{\tau \in \mathcal{T}_t}$ is Q-uniformly integrable.*

Proof. Note first that $\mathcal{M}$ is stable under concatenation, i.e. with Z^1, $Z^2 \in \mathcal{M}$ (here we identify density processes with their values at time T), and for every $\tau \in \mathcal{T}_t$, we have that

$$Z' := Z^1 \mathbb{I}_{[0,\tau[} + Z^2 \frac{Z^1_\tau}{Z^2_\tau} \mathbb{I}_{[\tau,T]} \in \mathcal{M}.$$

We claim that this implies for all $Q \in \mathcal{M}$ with Z, $\widetilde{Z}$ as above that

$$\log \widetilde{Z}^0 \leq \log \widetilde{Z}. \tag{7.10}$$

Indeed, if the claim were false, then there exists $\varepsilon > 0$ such that for the stopping time

$$\tau := \inf \Big\{ t \,\Big|\, \log \widetilde{Z}^0_t \geq \log \widetilde{Z}_t + \varepsilon \Big\} \wedge T$$

we would have $P(\tau < T) > 0$. Consider the concatenated density process

$$Z' := Z^0 \mathbb{I}_{[0,\tau[} + Z \frac{Z^0_\tau}{Z_\tau} \mathbb{I}_{[\tau,T]} \in \mathcal{M}.$$

By conditioning on $\mathcal{F}_\tau$ inside the expectation, and by the definition of τ, we obtain

$$E\left[Z'_T \log Z'_T \mathbb{I}_{\tau < T} \right]$$

$$= E\left[\left\{ Z^0_\tau \frac{Z_T}{Z_\tau} \left(\log Z^0_\tau + \log \frac{Z_T}{Z_\tau} \right) \right\} \mathbb{I}_{\tau < T} \right]$$

$$= E\left[Z^0_\tau \log Z^0_\tau \mathbb{I}_{\tau < T} \right] + E\left[Z^0_\tau \log \widetilde{Z}_\tau \mathbb{I}_{\tau < T} \right]$$

$$\leq E\left[Z^0_\tau \log Z^0_\tau \mathbb{I}_{\tau < T} \right] + E\left[Z^0_\tau \log \widetilde{Z}^0_\tau \mathbb{I}_{\tau < T} - \varepsilon E\left[Z^0_\tau \mathbb{I}_{\tau < T} \right] \right]$$

$$= E\left[Z^0_\tau \log Z^0_\tau \mathbb{I}_{\tau < T} \right] + E\left[Z^0_T \log \frac{Z^0_T}{Z^0_\tau} \mathbb{I}_{\tau < T} \right] - \varepsilon E\left[Z^0_\tau \mathbb{I}_{\tau < T} \right]$$

$$= E\left[Z^0_T \log Z^0_T \mathbb{I}_{\tau < T} \right] - \varepsilon E\left[Z^0_\tau \mathbb{I}_{\tau < T} \right].$$

As $Z'_\tau = Z^0_\tau$ on the set $\{\tau = T\}$ and $Z^0_\tau > 0$ on $\{\tau < T\}$, we get

$$E\left[Z'_T \log Z'_T\right] < E\left[Z^0_T \log Z^0_T\right],$$

which contradicts the minimality of Z^0_T. This proves the claim (7.10). Moreover, for every $\tau \in \mathcal{T}_t$ and $N > 0$ we have by Lemma 7.16

$$
\begin{aligned}
Q\left(\log \widetilde{Z}_\tau \geq N\right) &\leq \frac{1}{N} E_Q\left[\log \widetilde{Z}_\tau\right] \\
&= \frac{1}{N} E_Q\left[\log \frac{Z_T}{Z_\tau}\right] \\
&\leq \frac{1}{N} E\left[Z_T \log Z_T\right].
\end{aligned}
$$

Hence, it follows by (7.10) that

$$
\begin{aligned}
E_Q\left[\log \widetilde{Z}^0_\tau \mathbb{I}_{\log \widetilde{Z}^0_\tau \geq N}\right] &\leq E_Q\left[\log \widetilde{Z}_\tau \mathbb{I}_{\log \widetilde{Z}_\tau \geq N}\right] \\
&\leq E\left[(Z_T \log Z_T)\, \mathbb{I}_{\log \widetilde{Z}_\tau \geq N}\right],
\end{aligned}
$$

and therefore $\left\{\log \widetilde{Z}^0_\tau\right\}_{\tau \in \mathcal{T}_t}$ is Q-uniformly integrable. $\square$

Proposition 7.21. *Barron's inequality.* *Let Z be a density process. Then*

$$E\left[Z_T \sup_{0 \leq t \leq T} |\log Z_t|\right] \leq 2 + e + eE\left[Z_T \log Z_T\right].$$

Proof. See Kabanov & Stricker (2002), Proposition A.2. $\square$

We set now

$$\log \widehat{Z}^0_t := E_{Q^0}\left[\log Z^0_T \,\middle|\, \mathcal{F}_t\right]$$

and note that

$$\log \widehat{Z}^0 = \log \widetilde{Z}^0 + \log Z^0. \tag{7.11}$$

Theorem 7.22. *Entropic duality.* *We have*

$$\inf_{\vartheta \in \Theta} E\left[\exp\left(-\alpha \int_0^T \vartheta_t \, dS_t\right)\right] = \exp\left(-H\left(Q^0, P\right)\right) \tag{7.12}$$

and the inf is attained by $-\frac{1}{\alpha}\vartheta^0 \in \Theta$.

Proof. We will show that ϑ^0 is in Θ. The same calculation as in the proof of Proposition 7.19 then yields that the inf in (7.12) is attained by $-\frac{1}{\alpha}\vartheta^0 \in \Theta$.

Let $Q \in \mathcal{M}^f$. First note that without loss of generality we can assume that Q is equivalent to P: simply replace Q by $(Q + Q^0)/2$ if necessary. According to Lemma 7.11(i), we have that

$$H(Q, P) \geq H(Q, Q^0) + H(Q^0, P)$$

which implies that $H(Q, Q^0)$ is finite since $Q \in \mathcal{M}^f$. Denoting by Z^1 the density process of Q^0 with respect to Q and by Z^2 the density process of Q with respect to P, it follows by Barron's inequality that for both $i = 1, 2$ we get

$$\sup_{0 \leq t \leq T} \left|\log Z_t^i\right| \in L^1(Q).$$

Consequently, denoting by Z^0 the density process of Q^0 with respect to P, it follows readily from $Z^0 = Z^1 Z^2$ and the triangle inequality that

$$\sup_{0 \leq t \leq T} \left|\log Z_t^0\right| \in L^1(Q). \tag{7.13}$$

Hence the family $\left\{\log Z_\tau^0\right\}_{\tau \in \mathcal{T}_t}$ is Q-uniformly integrable. By Lemma 7.20 we have that $\left\{\log \tilde{Z}_\tau^0\right\}_{\tau \in \mathcal{T}_t}$ is Q-u.i. so it follows from (7.11) that $\left\{\log \hat{Z}_\tau^0\right\}_{\tau \in \mathcal{T}_t}$ is Q-u.i. as well. By the density representation (7.8), this is equivalent to the uniform integrability of $\left\{\int_0^\tau \vartheta_t^0 \, dS_t\right\}_{\tau \in \mathcal{T}_t}$, and therefore $\int \vartheta^0 \, dS$ is a true Q-martingale by Chou *et al.* (1980). □

Corollary 7.23. *The integrand ϑ^0 from the density representation (7.8) of Q^0 is in Θ.*

To extend the duality to cover the case when a claim B has been sold we impose the following standing assumption:

$$\boxed{B \text{ is a bounded claim.}}$$

This condition can be relaxed, see Becherer (2003) as well as Delbaen *et al.* (2002). We impose it here to reduce the need for extra integrability conditions with respect to a whole family of measures in the section about hedging.

Definition 7.24. We define a probability measure P^α equivalent to P by

$$\frac{dP^\alpha}{dP} = \exp\left(c_\alpha + \alpha B\right),$$

where $\exp(c_\alpha)$ is a normalising constant.

This change of measure will allow us to prove a duality involving B by reducing the problem to the already established situation of duality without claim. As B is bounded, the space $\mathcal{M}^f(P)$ equals $\mathcal{M}^f(P^\alpha)$ so that we will write simply $\mathcal{M}^f$ from now on. Therefore, with the convention that P corresponds to $\alpha = 0$, the space Θ of admissible strategies does not depend on α. For any bounded claim B and $Q \in \mathcal{M}^f$

$$H(Q, P) = E_Q\left[\log\frac{dQ}{dP^\alpha} + \log\frac{dP^\alpha}{dP}\right] \tag{7.14}$$

$$= E_Q\left[\log\frac{dQ}{dP^\alpha} + c_\alpha + \alpha B\right]$$

$$= H(Q, P^\alpha) + c_\alpha + E_Q\left[\alpha B\right].$$

By our assumption that $\mathcal{M}^f \cap \mathcal{M}^e \neq \varnothing$, there exists a unique element of $\mathcal{M}^f \cap \mathcal{M}^e$ which minimises $H(Q, P^\alpha)$ over all $Q \in \mathcal{M}^f$. This measure will be denoted by Q^α and called the minimal entropy martingale measure relative to P^α. Theorem 7.14 and Corollary 7.23 imply that its density can be written as

$$\frac{dQ^\alpha}{dP^\alpha} = \exp\left(c_\alpha^Q + \int_0^T \vartheta_t^\alpha\, dS_t\right)$$

for some constant c_α^Q and $\vartheta^\alpha \in \Theta$ such that $\int \vartheta^\alpha\, dS$ is a Q^α-martingale.

Theorem 7.25. *Entropic duality with claim.* *We have*

$$\inf_{\vartheta \in \Theta} E\left[\exp\left(-\alpha\left(\int_0^T \vartheta_t\, dS_t - B\right)\right)\right]$$

$$= \exp\left(-\inf_{Q \in \mathcal{M}^f}\left(H(Q, P) - E_Q\left[\alpha B\right]\right)\right)$$

and the inf are attained by $-\frac{1}{\alpha}\vartheta^\alpha \in \Theta$ *and* $Q^\alpha \in \mathcal{M}^f \cap \mathcal{M}^e \neq \varnothing$ *respectively.*

Proof. By changing the measure from P to P^α we can transform the l.h.s. of the equation as follows:

$$\inf_{\vartheta \in \Theta} E \left[\exp \left(-\alpha \left(\int_0^T \vartheta_t \, dS_t - B \right) \right) \right]$$

$$= e^{-c_\alpha} \inf_{\vartheta \in \Theta} E_{P^\alpha} \left[\exp \left(-\alpha \int_0^T \vartheta_t \, dS_t \right) \right],$$

whereas, from equation (7.14) and the definition of Q^α, the r.h.s. reads

$$\exp \left(- \inf_{Q \in \mathcal{M}^f} \left(H\left(Q, P\right) - E_Q \left[\alpha B \right] \right) \right)$$

$$= e^{-c_\alpha} \exp \left(- \inf_{Q \in \mathcal{M}^f} H\left(Q, P^\alpha\right) \right)$$

$$= e^{-c_\alpha} \exp \left(-H\left(Q^\alpha, P^\alpha\right) \right).$$

The result now follows directly from Theorem 7.22. $\square$

For the sake of clarity, we reformulate Theorem 7.25 in terms of the utility function $U_\alpha\left(x\right) = 1 - \exp\left(-\alpha x\right)$.

Corollary 7.26. *For any $c \in \mathbb{R}$, we have*

$$\sup_{\vartheta \in \Theta} E \left[U_\alpha \left(c + \int_0^T \vartheta_t \, dS_t - B \right) \right]$$

$$= U_\alpha \left(\inf_{Q \in \mathcal{M}^f} \left(\frac{1}{\alpha} H\left(Q, P\right) + c - E_Q\left[B\right] \right) \right).$$

As an application, we look at to which extent additional information can be exploited by an agent. In general, enlarging the filtration $\mathbb{F}^S$ generated by the price process can induce arbitrage opportunities when using strategies adapted to the larger filtration. Here we are looking at a less severe situation, and study the relationship between the immersion property and exponential utility maximisation via the duality to entropy minimisation. We denote the minimal entropy martingale measure relative to some filtration $\mathbb{H}$ with $Q^E\left(\mathbb{H}\right)$, its density process with $Z^E\left(\mathbb{H}\right)$.

Proposition 7.27. *Immersion and utility maximisation.* *Let $\mathbb{F} \subset \mathbb{G}$ be two filtrations, and let S be some $\mathbb{F}$-adapted process such that there exists a minimal entropy martingale measure $Q^E\left(\mathbb{G}\right)$ (and then $Q^E\left(\mathbb{F}\right)$ exists as well). If $\mathbb{F}$ is immersed in $\mathbb{G}$, then $Z^E\left(\mathbb{F}\right) = Z^E\left(\mathbb{G}\right)$.*

Proof. A martingale measure Q with respect to $\mathbb{G}$ induces a martingale measure $\widehat{Q}$ with respect to $\mathbb{F}$ via

$$\frac{d\widehat{Q}}{dP} = E\left[\frac{dQ}{dP}\bigg|\mathcal{F}_T\right].$$

By the conditional Jensen's inequality, we have $H\left(\widehat{Q}, P\right) \leq H(Q, P)$. Applying this to $Q^E(\mathbb{G})$ implies that $\widehat{Q^E(\mathbb{G})}$ has finite relative entropy, hence there is some martingale measure with respect to $\mathbb{F}$ which has finite relative entropy. Therefore $Q^E(\mathbb{F})$ exists uniquely. As $Z^E(\mathbb{F})\,S$ is a local $(P, \mathbb{F})$-martingale, it is by immersion also a local $(P, \mathbb{G})$-martingale. Hence $Z^E(\mathbb{F})$ is the density process of a martingale measure with respect to $\mathbb{G}$, with terminal value $dQ^E(\mathbb{F})/dP$. As we have

$$H\left(Q^E(\mathbb{F}), P\right) \leq H\left(\widehat{Q^E(\mathbb{G})}, P\right) \leq H\left(Q^E(\mathbb{G}), P\right),$$

the result follows by uniqueness of $Q^E(\mathbb{G})$. $\qquad\square$

As a consequence, if $\mathbb{F}$ is immersed in $\mathbb{G}$, then no additional expected exponential utility can be gained by employing $\mathbb{G}$-predictable instead of $\mathbb{F}$-predictable strategies.

7.4 Properties of the utility indifference price

Let us now return to discussing the utility indifference price $\pi(B; \alpha)$ introduced in Definition 7.1. Equation (7.2) reads as

$$\sup_{\vartheta \in \Theta} E\left[U_\alpha\left(c + \int_0^T \vartheta_t\, dS_t\right)\right]$$

$$= \sup_{\vartheta \in \Theta} E\left[U_\alpha\left(c + \pi(B; \alpha) + \int_0^T \vartheta_t\, dS_t - B\right)\right].$$

Alternatively, an equivalent dual formulation follows from Corollary 7.26 and the monotonicity and multiplicative property of the exponential function,

$$\inf_{Q \in \mathcal{M}^f} \frac{1}{\alpha} H(Q, P) = \inf_{Q \in \mathcal{M}^f}\left(\frac{1}{\alpha} H(Q, P) + \pi(B; \alpha) - E_Q[B]\right).$$

Under the standing assumptions $\pi(B; \alpha)$ can be expressed as

$$\pi(B; \alpha) = \sup_{Q \in \mathcal{M}^f}\left(E_Q[B] - \frac{1}{\alpha}\left(H(Q, P) - H(Q^0, P)\right)\right). \qquad (7.15)$$

This representation allows us to deduce the following basic properties of the exponential utility indifference price.

Corollary 7.28. *The utility indifference price $\pi(B;\alpha)$ satisfies the following properties:*

(i) $\pi(B;\alpha)$ *does not depend on the level of wealth c*

(ii) $\alpha \longmapsto \pi(B;\alpha)$ *is increasing in α*

(iii) $\pi(\beta B;\alpha) = \beta\pi(B;\beta\alpha)$ *for $\beta > 0$*

(iv) $\pi(B+c;\alpha) = \pi(B;\alpha) + c$ *for $c \in \mathbb{R}$*

(v) $\pi(B_1;\alpha) \leq \pi(B_2;\alpha)$ *if $B_1 \leq B_2$*

(vi) $\pi(\lambda B_1 + (1-\lambda)B_2;\alpha) \leq \lambda\pi(B_1;\alpha) + (1-\lambda)\pi(B_2;\alpha)$ *for any $\lambda \in [0,1]$*

(vii) *For an attainable claim $B = b + \int_0^T \vartheta_t\,dS_t$ where $\vartheta \in \Theta$, we have $\pi\left(b + \int_0^T \vartheta_t\,dS_t;\alpha\right) = b$.*

To prove (vii), notice that $\vartheta \in \Theta$ implies that $\int \vartheta\,dS$ is a Q-martingale for all $Q \in \mathcal{M}^f$ and therefore does not depend on Q since $E_Q[B] = b$ for all $Q \in \mathcal{M}^f$, so the result follows from (7.15).

Remark. The properties translation invariance (iv), monotonicity (v) and convexity (vi) imply that the mapping $\varrho : B \longmapsto \pi(-B;\alpha)$ satisfies all axioms required for it to constitute a **convex measure of risk** on the set of bounded claims.

The next proposition shows that as the risk-aversion parameter α tends to infinity, the utility indifference price tends to the super-replication price. Whereas, as $\alpha \downarrow 0$ we recover a linear pricing rule, under the minimal entropy martingale measure Q^0.

Proposition 7.29. *Risk-aversion asymptotics.* *We have*

$$(i) \quad \lim_{\alpha\uparrow\infty} \pi(B;\alpha) = \sup_{Q\in\mathcal{M}^e} E_Q[B]$$

$$(ii) \quad \lim_{\alpha\downarrow 0} \pi(B;\alpha) = E_{Q^0}[B].$$

Proof. ad (i): By monotonicity of $\pi(B;\alpha)$ in α (property (iv)) we get from (7.15)

$$\lim_{\alpha\uparrow\infty}\pi(B;\alpha) = \sup_{\alpha>0}\sup_{Q\in\mathcal{M}^f}\left(E_Q[B] - \frac{1}{\alpha}\left(H(Q,P) - H(Q^0,P)\right)\right)$$

$$= \sup_{Q\in\mathcal{M}^f}\sup_{\alpha>0}\left(E_Q[B] - \frac{1}{\alpha}\left(H(Q,P) - H(Q^0,P)\right)\right)$$

$$= \sup_{Q\in\mathcal{M}^f} E_Q[B].$$

The result follows since $\mathcal{M}^f$ (identified via its densities with a subset of $L^1(P)$) is $L^1(P)$-dense in $\mathcal{M}^e$, see Kabanov & Stricker (2001).

ad (ii): From (7.15) and the monotonicity of $\pi(B;\alpha)$ in α it follows that $\pi(B;\alpha) \geq E_{Q^0}[B]$ for all α and that $\pi(B;\alpha)$ decreases as α decreases. Therefore,

$$E_{Q^0}[B] \leq \lim_{\alpha\downarrow 0}\pi(B;\alpha).$$

To show the reverse inequality, observe that substituting P with Q^0 in Lemma 7.16 we obtain that for all $Q\in\mathcal{M}^f$

$$\alpha E_Q[B] \leq E_{Q^0}\left[e^{\alpha B}\right] + H(Q,Q^0) - 1.$$

It follows from (7.4) that

$$-\left(H(Q,P) - H(Q^0,P)\right) \leq -H(Q,Q^0),$$

and hence

$$E_Q[B] - \frac{1}{\alpha}\left(H(Q,P) - H(Q^0,P)\right) \leq E_{Q^0}\left[\frac{e^{\alpha B} - 1}{\alpha}\right].$$

Thus, taking the supremum over all $Q\in\mathcal{M}^f$ we obtain

$$\pi(B;\alpha) \leq E_{Q^0}\left[\frac{e^{\alpha B} - 1}{\alpha}\right].$$

Therefore, Lebesgue's dominated convergence theorem gives

$$\lim_{\alpha\downarrow 0}\pi(B;\alpha) \leq \lim_{\alpha\downarrow 0}E_{Q^0}\left[\frac{e^{\alpha B} - 1}{\alpha}\right] = E_{Q^0}[B].$$

□

Example 7.30. Merton's problem. We consider a risky asset with discounted continuous price process

$$S = M + \int \lambda \, d\,[M].$$

Assume that

$$K = \int \lambda^2 \, d\,[M] < \infty;$$

K is deterministic; and that there is a unique martingale measure Q for S. Hence

$$\frac{dQ}{dP} = \mathcal{E}\left(-\int \lambda \, dM\right)_T.$$

A prime example of when this framework holds is Samuelson's model where S is geometric Brownian motion. Merton's problem, with the addition of a bounded claim B, can be stated as

$$\inf_{\vartheta \in \Theta} E\left[\exp\left(\alpha\left(B - \int_0^T \vartheta_t \, dS_t\right)\right)\right].$$

As space of admissible strategies we choose

$$\Theta = \left\{\vartheta \,\middle|\, \int \vartheta \, dS \text{ is a } Q-\text{martingale}\right\}.$$

Notice that $\lambda \in \Theta$ because

$$E_Q\left[\left[\int \lambda \, dS\right]_T\right] = E_Q\,[K_T] = K_T < \infty$$

implies that $\int \lambda \, dS$ is a Q-martingale. The market is complete by the second fundamental theorem, so there exist some $\vartheta^B \in \Theta$ such that

$$B = c + \int_0^T \vartheta_t^B \, dS_t$$

where $c \in \mathbb{R}$. For any $\vartheta \in \Theta$ set $\psi = \vartheta - \vartheta^B \in \Theta$ so that

$$\inf_{\vartheta \in \Theta} E\left[\exp\left(\alpha\left(B - \int_0^T \vartheta_t \, dS_t\right)\right)\right]$$

$$= \inf_{\psi \in \Theta} E\left[\exp\left(\alpha\left(c - \int_0^T \psi_t \, dS_t\right)\right)\right].$$

By the duality result in Theorem 7.25 this infimum is attained and

$$
\min_{\psi \in \Theta} E \left[\exp \left(-\alpha \int_0^T \psi_t \, dS_t \right) \right]
$$

$$
= \exp \left(-E \left[\frac{dQ}{dP} \log \frac{dQ}{dP} \right] \right)
$$

$$
= \exp \left(-E_Q \left[-\int_0^T \lambda_t \, dM_t - \frac{1}{2} \int_0^T \lambda_t^2 \, d\,[M]_t \right] \right)
$$

$$
= \exp \left(-E_Q \left[-\int_0^T \lambda_t \, dS_t + \frac{1}{2} \int_0^T \lambda_t^2 \, d\,[M]_t \right] \right)
$$

$$
= \exp \left(-\frac{1}{2} K_T \right).
$$

The minimum is attained by $\psi = \frac{1}{\alpha}\lambda$ since

$$
E \left[\exp \left(-\int_0^T \lambda_t \, dS_t \right) \right]
$$

$$
= E \left[\exp \left(-\int_0^T \lambda_t \, dM_t - \frac{1}{2} \int_0^T \lambda_t^2 \, d\,[M]_t - \frac{1}{2} K_T \right) \right]
$$

$$
= E_Q \left[\exp \left(-\frac{1}{2} K_T \right) \right]
$$

$$
= \exp \left(-\frac{1}{2} K_T \right).
$$

Therefore, the optimal trading strategy for exponential utility with risk aversion parameter α is

$$
\psi^{opt} = \frac{1}{\alpha}\lambda.
$$

Moreover, the optimal strategy for the utility maximisation problem involving the claim is

$$
\vartheta^{opt} = \vartheta^B + \frac{1}{\alpha}\lambda \to \vartheta^B \quad \text{as } \alpha \to \infty.
$$

Hence the optimal strategy is the sum of the replicating strategy ϑ^B and the optimal investment strategy $\frac{1}{\alpha}\lambda$.

7.5 Entropic hedging

In this section the entropic hedging strategy for a bounded claim B will be introduced as suitable limit of utility indifference hedging strategies for vanishing risk-aversion, and related to the KW decomposition of B under the entropy measure Q^0.

Our standing assumptions are still in force, hence S is locally bounded and $\mathcal{M}^f \cap \mathcal{M}^e \neq \varnothing$. Throughout, we take the space Θ introduced previously as space of admissible strategies. By Theorem 7.14 and Corollary 7.23, there exists a constant c_0 and a unique $\vartheta^0 \in \Theta$ such that

$$\frac{dQ^0}{dP} = \exp\left(c_0 + \int_0^T \vartheta_t^0 \, dS_t\right). \tag{7.16}$$

Moreover, the strategy $\theta = -\frac{1}{\alpha}\vartheta^0$ maximises the expected utility in the pure investment problem as in the l.h.s. of (7.12). Recall that the density of the probability measure P^α with respect to P is

$$\frac{dP^\alpha}{dP} = \exp\left(c_\alpha + \alpha B\right).$$

The boundedness of B implies that $\mathcal{M}^f(P)$ equals $\mathcal{M}^f(P^\alpha)$ so that we will write simply $\mathcal{M}^f$ from now on. We even have $\mathcal{M}^f(P) \subset \mathcal{M}^f(Q^0)$ which follows readily from the equation

$$H\left(Q, Q^0\right) = H\left(Q, P\right) - H\left(Q^0, P\right) \quad \forall Q \in \mathcal{M}^f(P).$$

Moreover, since $\mathcal{M}^f \cap \mathcal{M}^e \neq \varnothing$, there exists uniquely a minimal entropy martingale measure Q^α relative to P^α. Its density can be uniquely written as

$$\frac{dQ^\alpha}{dP^\alpha} = \exp\left(c_\alpha^Q + \int_0^T \vartheta_t^\alpha \, dS_t\right)$$

for some $\vartheta^\alpha \in \Theta$. Moreover, the strategy $\theta = -\frac{1}{\alpha}\vartheta^\alpha$ solves the utility maximisation problem (recall $U_\alpha(x) = 1 - \exp(-\alpha x)$)

$$\max_{\theta \in \Theta} E\left[U_\alpha\left(\pi + \int_0^T \theta_t \, dS_t - B\right)\right] \tag{7.17}$$

of an investor who has sold contingent claim B at time 0 for an arbitrary price π.

The density process of Q^α with respect to Q^0 will be denoted by Z^α. It then follows from the formulae for the densities of Q^0 and P^α with respect to P that there exist a constant d_α and a $\vartheta^\alpha \in \Theta$ such that

$$Z_T^\alpha = \frac{dQ^\alpha}{dP^\alpha} \times \frac{dP^\alpha}{dP} \times \frac{dP}{dQ^0} \tag{7.18}$$

$$= \exp\left(d_\alpha + \alpha B + \int_0^T \left(\vartheta_t^\alpha - \vartheta_t^0\right) dS_t\right).$$

To motivate our main convergence result, we first consider the case where the claim B is attainable so that there exists a strategy $\vartheta^B \in \Theta$ such that

$$B = E_{Q^0}[B] + \int_0^T \vartheta_t^B \, dS_t.$$

Proposition 7.31. *If B is attainable, we have for any $\alpha > 0$ that*

$$\frac{1}{\alpha}\int \left(\vartheta^0 - \vartheta^\alpha\right) dS = \int \vartheta^B \, dS.$$

Proof. Using the attainability, we write

$$\log \frac{dQ^\alpha}{dP} = \log \frac{dP^\alpha}{dP} + \log \frac{dQ^\alpha}{dP^\alpha}$$

$$= d_\alpha + \alpha B + \int_0^T \vartheta_t^\alpha \, dS_t$$

$$= d_\alpha + \alpha E_{Q^0}[B] + \int_0^T \left(\alpha\vartheta_t^B + \vartheta_t^\alpha\right) dS_t.$$

As $\alpha\vartheta^B + \vartheta^\alpha \in \Theta$, Proposition 7.15 implies that $dQ^\alpha/dP = Z_T^0$, and hence we can identify $\int \vartheta^0 \, dS$ with $\int \left(\alpha\vartheta^B + \vartheta^\alpha\right) dS$. $\quad\square$

This result shows that when B is attainable, the value process associated with the strategy $\frac{1}{\alpha}\left(\vartheta^0 - \vartheta^\alpha\right)$ coincides with the value process from using the unique replicating strategy ϑ^B. The strategy $\frac{1}{\alpha}(\vartheta^0 - \vartheta^\alpha)$ represents the adjustment to the investor's portfolio when the contingent claim B is sold.

Definition 7.32. $\phi^\alpha := \frac{1}{\alpha}(\vartheta^0 - \vartheta^\alpha)$ is called the **exponential utility-indifference hedging strategy**. Denote by $\pi^\alpha = \pi(B;\alpha)$ the indifference price as defined in Definition 7.1. π^α is unique because the exponential function is strictly increasing. Our goal is to approximate ϕ^α and π^α for small risk aversion α in the case where the bounded claim B is not necessarily attainable. Consider for this purpose the KW decomposition of B under Q^0:

$$\widetilde{B_t^0} := E_{Q^0}[B|\mathcal{F}_t] = E_{Q^0}[B] + \int_0^t \vartheta_u^B \, dS_u + L_t \qquad (7.19)$$

for a predictable process $\vartheta^B \in L^2(S)$ and a local Q^0-martingale L strongly orthogonal to S. That is $[S, L]$ is a local Q^0-martingale and hence $\langle S, L \rangle = 0$ where the angle bracket process is calculated with respect to Q^0. As discussed in Chapter 6, the mean-variance price equals $c^B := E_{Q^0}[B]$, and ϑ^B represents the optimal hedge of B relative to a mean square distance under the entropy measure Q^0. We denote the expected squared hedging error by

$$\epsilon^2 := E_{Q^0}[L_T^2].$$

The proof of the next theorem is primarily based on Taylor expansions but requires some delicate estimates, for which we refer to Kallsen & Rheinländer (2011). In the sequel, we set

$$X_t^\alpha = E_{Q^\alpha}[B|\mathcal{F}_t] - \frac{1}{\alpha} \int_0^t \left(\vartheta_u^0 - \vartheta_u^\alpha \right) dS_u,$$

and write $f(\alpha) \sim o(\alpha)$ when a real-valued function f is such that $f(\alpha)/\alpha \to 0$ as $\alpha \to 0$. Similarly, $f(\alpha) \sim O(\alpha)$ when $f(\alpha)/\alpha$ is bounded. For processes this is to be understood a.s. pathwise.

Theorem 7.33. *Let S be a continuous semi-martingale. We have*

$$\|\vartheta^B - \phi^\alpha\|_{L^2(S)} \xrightarrow{\alpha \to 0} 0,$$

$$\pi^\alpha = c^B + \frac{\alpha}{2}\epsilon^2 + o(\alpha), \qquad (7.20)$$

where

$$\|\theta\|_{L^2(S)} := \sqrt{E_{Q^0}\left[\int_0^T \theta_t^2 \, d[S]_t\right]}.$$

In particular, this implies

$$\sup_{t \in [0,T]} \left| \int_0^t \vartheta_u^B \, dS_u - \int_0^t \phi_u^\alpha \, dS_u \right| \to 0 \quad in \ L^2(Q^0).$$

Proof. (sketch) Recall that the inf in the exponential hedging problem

$$\frac{1}{\alpha} \inf_{\vartheta \in \Theta} E \left[\exp \left(c_0 + \alpha B - \int_0^T \left(\alpha \vartheta_t - \vartheta_t^0 \right) dS_t \right) \right]$$

is uniquely attained by $\phi^\alpha = \frac{1}{\alpha} \left(\vartheta^0 - \vartheta^\alpha \right)$. We proceed by using the exponential series, denoting the remainder term by

$$r(x) = \left(\frac{1}{6} x + \frac{1}{24} x^2 + \ldots \right).$$

For ease of notation set $X^\alpha = X_T^\alpha$. By a change of measure according to (7.16) and the minimality of ϕ^α we get that

$$\frac{1}{\alpha} E \left[\exp \left(c_0 + \alpha B - \int_0^T \left(\alpha \phi_t^\alpha - \vartheta_t^0 \right) dS_t \right) - 1 \right]$$

$$= \frac{1}{\alpha} E_{Q^0} \left[\exp \left(\alpha B - \alpha \int_0^T \phi_t^\alpha \, dS_t \right) - 1 \right]$$

$$= E_{Q^0} [B] + \frac{\alpha}{2} E_{Q^0} \left[\left(B - \int_0^T \phi_t^\alpha \, dS_t \right)^2 \right] + \alpha E_{Q^0} \left[r \left(\alpha X^\alpha \right) \left(X^\alpha \right)^2 \right]$$

$$\leq E_{Q^0} [B] + \frac{\alpha}{2} E_{Q^0} \left[\left(B - \int_0^T \vartheta_t^B \, dS_t \right)^2 \right] + \alpha E_{Q^0} \left[r \left(\alpha \tilde{X} \right) \tilde{X}^2 \right]$$

where $\tilde{X} = B - \int_0^T \vartheta_t^B \, dS_t$ does not depend on α. We obtain

$$E_{Q^0} \left[\left(B - \int_0^T \phi_t^\alpha \, dS_t \right)^2 \right] \leq E_{Q^0} \left[\left(B - \int_0^T \vartheta_t^B \, dS_t \right)^2 \right]$$

$$+ \left| E_{Q^0} \left[r(\alpha X^\alpha)(X^\alpha)^2 \right] \right| + \left| E_{Q^0} \left[r(\alpha \tilde{X}) \tilde{X}^2 \right] \right|. \tag{7.21}$$

We can estimate the remainder term by

$$\left| r\left(\alpha x \right) x^2 \right| \leq r \left(\alpha \left| x \right| \right) x^2 \leq \left(e^{\alpha |x|} - 1 \right) x^2 \leq e^{(\alpha + \varepsilon)|x|} - 1 + O(\alpha)$$

for any $\varepsilon > 0$. Therefore

$$\left| E_{Q^0} \left[r \left(\alpha X^\alpha \right) \left(X^\alpha \right)^2 \right] \right| \leq E_{Q^0} \left[\left(e^{(\alpha + \varepsilon)|X^\alpha|} - 1 \right) \right] + O(\alpha).$$

One can show that this tends to zero for $\alpha \to 0$. In addition, we have

$$r \left(\alpha \tilde{X} \right) \to 0 \qquad Q^0 - \text{a.s. for } \alpha \to 0.$$

Moreover, it can be shown that

$$\left| E_{Q^0} \left[r \left(\alpha \tilde{X} \right) \tilde{X}^2 \right] \right| \to 0.$$

We therefore get from (7.21) that

$$\lim_{\alpha \to 0} E_{Q^0} \left[\left(B - \int_0^T \phi_t^\alpha \, dS_t \right)^2 \right] \le E_{Q^0} \left[\left(B - \int_0^T \vartheta_t^B \, dS_t \right)^2 \right],$$

hence by the minimality of ϑ^B as optimal mean-variance hedge

$$\lim_{\alpha \to 0} E_{Q^0} \left[\left(B - \int_0^T \phi_t^\alpha \, dS_t \right)^2 \right] = E_{Q^0} \left[\left(B - \int_0^T \vartheta_t^B \, dS_t \right)^2 \right].$$

Since ϑ^B is minimal projection in $L^2(Q^0)$, we get by Pythagoras

$$\left\| B - \int_0^T \phi_t^\alpha \, dS_t \right\|_{L^2(Q^0)}^2 = \left\| B - \int_0^T \vartheta_t^B \, dS_t \right\|_{L^2(Q^0)}^2$$

$$+ \left\| \int_0^T \vartheta_t^B \, dS_t - \int_0^T \phi_t^\alpha \, dS_t \right\|_{L^2(Q^0)}^2,$$

and hence

$$\left\| \int_0^T \vartheta_t^B \, dS_t - \int_0^T \phi_t^\alpha \, dS_t \right\|_{L^2(Q^0)} \to 0.$$

The assertion follows since convergence of square-integrable martingales in $\mathcal{H}^2(Q^0)$ is equivalent to convergence of their final values in $L^2(Q^0)$.

The indifference condition (7.2) can be rewritten as

$$\frac{E \left[\exp \left(- \int_0^T \vartheta_t^0 \, dS_t \right) \right]}{e^{c_0}} = E \left[\exp \left(-\alpha \pi^\alpha - \int_0^T (\vartheta_t^0 - \alpha \phi_t^\alpha) \, dS_t - \alpha B \right) \right]$$

or rather

$$\frac{\exp(\alpha \pi^\alpha) - 1}{\alpha} = \frac{1}{\alpha} E \left[\exp \left(c_0 + \alpha B - \int_0^T (\alpha \phi_t^\alpha - \vartheta_t^0) \, dS_t \right) - 1 \right].$$

By Taylor expansion, the r.h.s. is of the form, up to terms of $o(\alpha)$,

$$E_{Q^0}[B] + \frac{\alpha}{2} E_{Q^0} \left[\left(B - \int_0^T \vartheta_t^B \, dS_t \right)^2 \right] = c^B + \frac{\alpha}{2} \left((c^B)^2 + \epsilon^2 \right),$$

which in turn yields

$$\exp(\alpha\pi^\alpha) = 1 + \alpha c^B + \frac{\alpha^2}{2}\left((c^B)^2 + \epsilon^2\right) + o(\alpha^2)$$

$$= \exp\left(\alpha c^B + \frac{\alpha^2}{2}\epsilon^2 + o(\alpha^2)\right).$$

Taking logarithms and dividing by α yields (7.20). □

7.6 Notes and further reading

An overview of utility indifference pricing is provided in Carmona (2008). Existence and uniqueness results regarding the minimal entropy martingale measure were obtained by Frittelli (2000), and the key representation for its density is based on Csiszár (1975) and Yor (1978). For the duality results and further ramifications we refer to the six-authors paper Delbaen *et al.* (2002), Kabanov & Stricker (2002), and Schachermayer (2003). The qualitative properties of the exponential utility indifference price were obtained by Becherer (2003). Duality for more general utility functions has been studied by Goll & Rüschendorff (2001), and is the subject of the monograph by Frittelli *et al.* (2011). In the case when the price process is not locally bounded, one can recover the duality with an extra penalisation term relating to extreme events, see Biagini & Frittelli (2008), and the utility indifference price in this setting has been studied in Biagini *et al.* (2011). The limiting price for vanishing risk-aversion has been identified in Stricker (2004), and the limiting hedge in various degrees of generality in Becherer (2006), Mania & Schweizer (2005) and Kallsen & Rheinländer (2011).

When statically replicating a general European payoff as in (5.4), ideally one would invest into a portfolio of vanilla options with a continuum of strikes. Ilhan & Sircar (2006) propose to approximate this portfolio using a limited number of options, and to hedge the remaining risk dynamically by employing an exponential utility indifference approach. Furthermore, the indifference approach has been applied to the hedging of volatility derivatives by Grasselli & Hurd (2007). An asymptotic expansion of the indifference price in the context of hedging of basis risk with partial information has been reached in Monoyios (2010).

Chapter 8

Hedging Constraints

In this chapter, pricing and hedging in the presence of trading constraints will be studied using utility indifference methods. When there are constraints on trading, a market is incomplete if it is not possible to execute the replicating strategy of an arbitrary claim. Even when each asset price is assumed to follow geometric Brownian motion, when some assets cannot be traded, in general, the market will be incomplete. When an asset cannot be directly traded, an imperfect substitute can be used to produce a partial hedge. Trading constraints leave a residual 'basis risk' due to a discrepancy between the traded and untraded asset price processes.

In contrast to the duality method introduced in the previous chapter, here the aim is to directly maximise the agents' utility from terminal wealth. The martingale optimality principle is used to link this problem to the solution of a quadratic backward stochastic differential equation (BSDE). In the absence of trading constraints, this BSDE is related to the density of the minimal entropy martingale measure.

8.1   Framework and preliminaries

Consider an agent with exponential utility from wealth at terminal time $T < \infty$, so that the utility function is

$$U_\alpha (x) = 1 - \exp (-\alpha x) \qquad \text{for } \alpha > 0.$$

Throughout this chapter we work on a filtered probability space $(\Omega, \mathcal{F}, \mathbb{F}, P)$ satisfying the usual conditions and such that all $\mathbb{F}$-martingales are continuous. Take an adapted n-dimensional martingale N with $[N^i, N^j] = 0$ for all $i \neq j$ describing the independent risk factors driving the financial market. The $\mathbb{R}^n$-valued price process of the assets studied takes the form

$S = \mathcal{E}(Y)$ where

$$dY = dM + \lambda' \, d[M] \, ,$$

for an $\mathbb{R}^n$-valued predictable process λ such that

$$\int_0^T \lambda_t' \, d[M]_t \, \lambda_t < \infty$$

and $M = \nu N$ where ν is an $\mathbb{R}^{n \times n}$-valued process with full-rank that determines the influence of the underlying risk factors on each of the assets. As a consequence of the GKW decomposition, Theorem 2.33, any square-integrable $\mathbb{F}$-martingale K can be expressed in the form

$$K = \int Z' \, dM + L$$

where $Z \in L^2(M)$ is an $\mathbb{R}^n$-valued predictable process and L is a square-integrable martingale strongly orthogonal to each component of M, that is $[M^i, L] = 0$ for each $1 \le i \le n$.

A **trading strategy** $\vartheta = (\vartheta^1, \ldots, \vartheta^n)'$ is a vector-valued process describing the number of units of each asset held and such that $\vartheta \in L(S)$ to ensure that the gains process $\int \vartheta' \, dS$ is well-defined. Given an initial wealth c, the wealth process associated with a strategy ϑ is

$$V^{(\vartheta)} = c + \int \vartheta' \, dS. \tag{8.1}$$

Even without additional constraints on the choice of trading strategy, it may not be possible to replicate all $\mathbb{F}$-martingales using this wealth process. The following definition of an admissible strategy both introduces the concept of trading constraints and also constrains us to use only arbitrage-free strategies.

Definition 8.1. A closed convex subset $K \subset \mathbb{R}^n$ containing the origin is referred to as the **strategy space**. A strategy ϑ is in the set of **admissible strategies** Θ_K if it satisfies the following three conditions:

(*i*) For each $t \in [0, T]$

$$\mathrm{diag}\,(S_t) \, \vartheta_t \in K$$

where $\mathrm{diag}(S)$ denotes the $\mathbb{R}^{n \times n}$-valued process with S on the lead diagonal and zero elsewhere;

(*ii*) The strategy satisfies

$$E\left[\int_0^T \left(\mathrm{diag}\,(S_t)\, \vartheta_t \right)' \, d[M]_t \left(\mathrm{diag}\,(S_t)\, \vartheta_t \right) \right] < \infty;$$

(*iii*) The family of random variables

$$\left\{ \exp\left(-\alpha V_\tau^{(\vartheta)}\right) \middle| \tau \leq T \text{ is a stopping time} \right\}$$

is uniformly integrable.

The second condition ensures that the stochastic integral in (8.1) is a square-integrable martingale and is sufficient to ensure that the strategy does not represent an arbitrage opportunity. Furthermore, under these assumptions, for each admissible strategy the resulting wealth process can be expressed in the form

$$V_t^{(\vartheta)} = V_T^{(\vartheta)} - \int_t^T \left(\operatorname{diag}(S_t)\,\vartheta_t\right)' d\,[M]_s\,\lambda_s - \int_t^T \left(\operatorname{diag}(S_t)\,\vartheta_t\right)' dM_s \quad (8.2)$$

and each attainable terminal wealth $V_T^{(\vartheta)}$ can be uniquely determined in terms of an initial wealth $c \in \mathbb{R}$ and an admissible strategy $\vartheta \in \Theta_K$. This section concludes with some examples of trading constraints that can be dealt with using the approach discussed in this chapter.

(A) Basis risk — It is not possible to trade in one of the underlying assets but another asset is available which is an imperfect substitute for the untradeable asset.

(B) Limits on short selling — There is a limit $s > 0$ on the amount of each risky asset that can be shorted, i.e.

$$K = [-s, \infty)^n$$

or alternatively, there is a limit $s > 0$ on the total number of assets which are short-sold so that

$$K = \left\{ x \in \mathbb{R}^n \,\middle|\, \sum_{i=1}^n x^i \mathbb{I}_{x^i < 0} \geq -s \right\}.$$

(C) Margin account — A proportion of $\delta \in (0,1]$ of the value of any position in the risky assets held must be set aside in a margin account. At each $t \in [0,T]$ this constraint requires that

$$V^{(\vartheta)} - \|\operatorname{diag}(S)\,\vartheta\| \geq \delta \|\operatorname{diag}(S)\,\vartheta\|$$

and thus the strategy space depends on the total wealth, namely:

$$K(t,\omega) = \left\{ x \in \mathbb{R}^n \,\middle|\, V_t^{(\vartheta)} \geq (1+\delta) \|\operatorname{diag}(S_t)\,x\| \right\}.$$

This type of constraint is common in futures markets.

In this final example the constraint set K depends on (t, ω) and is not convex and hence does not satisfy Definition 8.1. However, the methods described in this chapter can be extended to cover closed but not necessarily convex strategy spaces, as explained in the notes at the end of this chapter.

8.2   Dynamic utility indifference pricing

Consider an agent who has sold a bounded $\mathcal{F}_T$-measurable contingent claim denoted B. The maximum expected utility the agent can expect to achieve without having sold the claim B is

$$u(c; \alpha) = \sup_{\vartheta \in \Theta_K} E\left[-\exp\left(-\alpha\left(c + \int_0^T \vartheta_u' \, dS_u\right)\right)\right]. \qquad (8.3)$$

Whereas the maximum expected utility if the claim has been sold at price π is

$$u(c + \pi - B; \alpha) = \sup_{\vartheta \in \Theta_K} E\left[-\exp\left(-\alpha\left(c + \pi - B + \int_0^T \vartheta_u' \, dS_u\right)\right)\right]. \qquad (8.4)$$

As in the previous chapter, we define the utility indifference price as follows:

Definition 8.2. Fix a bounded claim B and initial wealth $c \in \mathbb{R}$. If a unique solution $\pi = \pi(B; \alpha)$ to the equation

$$u(c; \alpha) = u(c + \pi - B; \alpha) \qquad (8.5)$$

exists, it is called the **utility indifference price** for B.

Furthermore, assuming that the claim B has been sold for an arbitrary price π, we can also retrieve from the solutions to the utility maximisation problems (8.3) and (8.4) the adjustment to the trading strategy required to hedge the claim.

Definition 8.3. Fix π and let $\vartheta^0, \vartheta^* \in \Theta_K$ be the unique trading strategies which achieve the suprema in (8.3) and (8.4) respectively. Then the trading strategy associated with the difference between these two strategies

$$\vartheta^B = \vartheta^* - \vartheta^0$$

is the **utility indifference hedge** for B.

The definition of an admissible strategy in this chapter differs from that used in Chapter 7 so this hedge is not necessarily related to the minimal entropy measure. We discuss in more detail the connection between the two approaches in Section 8.6 below.

Consider the problem of finding a hedging strategy $\vartheta \in \Theta_K$ which maximises the wealth on the outstanding period $[t, T]$ given that the wealth at time t equals c,

$$u(t, c - B; \alpha) := \operatorname*{ess\ sup}_{\vartheta \in \Theta_K} E\left[-\exp\left(\alpha \left(B - c - \int_t^T \vartheta'_u \, dS_u \right) \right) \middle| \mathcal{F}_t \right].$$
(8.6)

This is a dynamic version of the utility maximisation problem (8.4) that can be used to extend Definition 8.5 to describe a utility indifference price process.

Definition 8.4. Fix a bounded claim B and suppose that for each $t \in [0, T]$ there exist finite random variables $u(t, c - B; \alpha)$ and $u(t, c; \alpha)$ defined via (8.6). If a process $\pi = \pi(B; \alpha)$ exists such that for each $t \in [0, T]$

$$\pi_t = \frac{1}{\alpha} \log\left(\frac{u(t, c - B; \alpha)}{u(t, c; \alpha)} \right)$$

then it is called the **utility indifference price process** for B.

8.3 Martingale optimality principle

The following theorem characterises the trading strategy which achieves the supremum in (8.4) and also (8.6). Without loss of generality we may take $c = 0$.

Theorem 8.5. *Martingale optimality principle. For a given bounded claim B there exists a cadlag process J^B such that for each $t \in [0, T]$*

$$J_t^B = \operatorname*{ess\ inf}_{\vartheta \in \Theta_K} E\left[\exp\left(\alpha \left(B - \int_t^T \vartheta'_u \, dS_u \right) \right) \middle| \mathcal{F}_t \right].$$
(8.7)

The process J^B is the largest process J with terminal value $J_T = e^{\alpha B}$ such that for each $\vartheta \in \Theta_K$ the process $J^{(\vartheta)}$ defined via

$$J^{(\vartheta)} := J \exp\left(-\alpha \int \vartheta' \, dS \right)$$
(8.8)

is a sub-martingale. Moreover, the following two statements are equivalent:

(i) *The strategy* $\vartheta^* \in \Theta_K$ *achieves the supremum in (8.6) for each* $t \in [0, T]$ *and*

$$u(t, -B; \alpha) = E\left[-\exp\left(\alpha\left(B - \int_t^T \vartheta_u^{*\prime}\, dS_u\right)\right)\middle|\mathcal{F}_t\right] = -J_t^B.$$

(ii) $\vartheta^* \in \Theta_K$ *is such that the process* $J^{(\vartheta^*)}$ *is a martingale.*

Proof.   Define a family of random variables indexed by $t \in [0, T]$ as

$$\tilde{J}_t^B := \operatorname{ess\,inf}_{\vartheta \in \Theta_K} E\left[\exp\left(\alpha\left(B - \int_t^T \vartheta_u^\prime\, dS_u\right)\right)\middle|\mathcal{F}_t\right]. \qquad (8.9)$$

This is not yet a stochastic process as each $\tilde{J}_t^B$ is only defined P-a.s. We denote by $\tilde{J}^B$ an $\mathbb{F}$-adapted process which is equal to this limit almost everywhere, and look for a right-continuous modification. Restrict Θ_K to the interval $[t, T]$ by setting

$$\Theta_K(t) := \{\vartheta \in \Theta_K \,|\, \vartheta_u = 0 \text{ for all } u \in [0, t]\},$$

and for $\vartheta \in \Theta_K(t)$ let

$$R(t, \vartheta) := E\left[\exp\left(\alpha\left(B - \int_t^T \vartheta_u^\prime\, dS_u\right)\right)\middle|\mathcal{F}_t\right],$$

so that

$$\tilde{J}_t^B = \operatorname{ess\,inf}_{\vartheta \in \Theta_K(t)} R(t, \vartheta).$$

Take any $\varphi^1, \varphi^2 \in \Theta_K(t)$, let $A := \{R(t, \varphi^2) \geq R(t, \varphi^1)\} \in \mathcal{F}_t$ and set $\varphi = \varphi^1 \mathbb{I}_A + \varphi^2 \mathbb{I}_{A^c}$, then

$$R(t, \varphi) = R(t, \varphi^1)\,\mathbb{I}_A + R(t, \varphi^2)\,\mathbb{I}_{A^c}.$$

This shows that for arbitrary $\varphi^1, \varphi^2 \in \Theta_K(t)$ there exists $\varphi \in \Theta_K(t)$ such that $R(t, \varphi) = R(t, \varphi^1) \wedge R(t, \varphi^2)$, so the set of random variables $\{R(t, \vartheta)\,|\, \vartheta \in \Theta_K(t)\}$ is stable with respect to taking the minimum. Following Neveu (1975), this implies the existence of a minimising sequence $(\varphi^p)_{p \geq 1}$ such that $R(t, \varphi^p) \geq R(t, \varphi^{p+1})$ for each $p \geq 1$ and

$$\tilde{J}_t^B = \lim_{p \to \infty} R(t, \varphi^p).$$

We claim that the process $\tilde{J}^{(\vartheta)}$ defined as

$$\tilde{J}^{(\vartheta)} := \tilde{J}^B \exp\left(-\alpha \int \vartheta^\prime\, dS\right)$$

is a sub-martingale for all $\vartheta \in \Theta_K$. To prove this, we take $s < t \leq T$ and $\vartheta \in \Theta_K$. Then by monotone convergence

$$E\left[\exp\left(-\alpha\int_s^t \vartheta_u' \, dS_u\right) \tilde{J}_t^B \bigg| \mathcal{F}_s\right]$$

$$= \lim_{p\to\infty} E\left[\exp\left(-\alpha\int_s^t \vartheta_u' \, dS_u\right) R(t, \varphi^p) \bigg| \mathcal{F}_s\right]$$

$$= \lim_{p\to\infty} E\left[\exp\left(-\alpha\int_s^t \vartheta_u' \, dS_u\right)\exp\left(\alpha\left(B - \int_t^T \varphi_u^{p'} \, dS_u\right)\right) \bigg| \mathcal{F}_s\right]$$

$$= \lim_{p\to\infty} E\left[\exp\left(\alpha\left(B - \int_s^T \tilde{\varphi}_u^{p'} \, dS_u\right)\right) \bigg| \mathcal{F}_s\right] \geq \tilde{J}_s^B$$

since the sequence $(\tilde{\varphi}^p)_{p\geq 1}$ such that

$$\tilde{\varphi}_u^p = \varphi_u^p \mathbb{I}_{(t,T]}(u) + \vartheta_u' \mathbb{I}_{(s,t]}(u) \in \Theta_K(s)$$

is not necessarily minimising. Multiplying both sides by $\exp\left(-\alpha\int_0^s \vartheta_u' \, dS_u\right)$ yields the claim.

Since $\mathbb{F}$ satisfies the usual conditions, we can find a cadlag modification $\tilde{J}_+^{(\vartheta)}$ of the sub-martingale $\tilde{J}^{(\vartheta)}$ which is also an $\mathbb{F}$-sub-martingale; see Revuz & Yor (1999), Theorem II.2.9. We then take the process J^B to be equal to $\tilde{J}_+^B = \tilde{J}_+^{(\vartheta)} \exp\left(\alpha\int_0^{\cdot} \vartheta_u' \, dS_u\right)$.

Finally we have to show that J^B is the largest process with the property stated in the theorem. Suppose that J is another process such that

$$J^{(\vartheta)} = J \exp\left(-\alpha\int \vartheta' \, dS\right)$$

is a sub-martingale for all $\vartheta \in \Theta_K$ with $J_T = e^{\alpha B}$. The sub-martingale property implies that

$$J_t^{(\vartheta)} \leq E\left[\exp\left(\alpha\left(B - \int_0^T \vartheta_u' \, dS_u\right)\right) \bigg| \mathcal{F}_t\right]$$

$$= \exp\left(-\alpha\int_0^t \vartheta_u' \, dS_u\right) R(t, \vartheta\mathbb{I}_{[t,T]}).$$

This shows that $J_t \leq R(t, \vartheta\mathbb{I}_{[t,T]})$, and taking the infimum over the set $\Theta_K(t)$ yields $J_t \leq J_t^B$.

To see that $(i) \Rightarrow (ii)$, notice that if ϑ^* achieves the essential supremum in (8.7) for each $t \in [0,T]$ then $J^{(\vartheta^*)}$ is a martingale since

$$J_t^{(\vartheta^*)} = J_t^B \exp\left(-\alpha \int_0^t \vartheta_u^{*\prime} \, dS_u\right)$$

$$= E\left[\exp\left(\alpha\left(B - \int_0^T \vartheta_u^{*\prime} \, dS_u\right)\right)\Big| \mathcal{F}_t\right].$$

To show that $(ii) \Rightarrow (i)$, suppose that $J^{(\vartheta^*)}$ is a martingale. As each element of the family of processes $\{\, J^{(\vartheta)} \,|\, \vartheta \in \Theta_K \,\}$ is a sub-martingale and $J_0^{(\vartheta)} = J_0^B$ for each $\vartheta \in \Theta_K$ we have

$$\exp\left(\alpha \int_0^t \vartheta_u^{\prime} \, dS_u\right) E\left[J_T^{(\vartheta)}\Big| \mathcal{F}_t\right] \geq J_t^B = \exp\left(\alpha \int_0^t \vartheta_u^{*\prime} \, dS_u\right) E\left[J_T^{(\vartheta^*)}\Big| \mathcal{F}_t\right].$$

From this we deduce that $u\,(t, -B; \alpha) = -J_t^B$. $\qquad\square$

As a consequence of the previous theorem it is possible to solve the utility optimisation problem (8.4) by identifying a family of processes, denoted $\{\, J^{(\vartheta)} \,|\, \vartheta \in \Theta_K \,\}$, with the following three properties:

(1) For each $\vartheta \in \Theta_K$ the terminal value of the process $J^{(\vartheta)}$ is

$$J_T^{(\vartheta)} = \exp\left(-\alpha\left(V_T^{(\vartheta)} - B\right)\right);$$

(2) The processes $\{\, J^{(\vartheta)} \,|\, \vartheta \in \Theta_K \,\}$ all have the same deterministic initial value, $J_0^{(\vartheta)} = J_0$ for all $\vartheta \in \Theta_K$;
(3) For each $\vartheta \in \Theta_K$ the process $J^{(\vartheta)}$ is a sub-martingale and there exists $\vartheta^* \in \Theta_K$ such that $J^{(\vartheta^*)}$ is a martingale.

Once a family of processes $\{\, J^{(\vartheta)} \,|\, \vartheta \in \Theta_K \,\}$ with these three properties has been found, the previous theorem implies that to solve the utility maximisation problem (8.4), we need only observe that

$$-E\left[J_T^{(\vartheta)}\right] = E\left[-\exp\left(-\alpha\left(V_T^{(\vartheta)} - B\right)\right)\right]$$

$$\leq -J_0 e^{-\alpha c} = E\left[-\exp\left(-\alpha\left(V_T^{(\vartheta^*)} - B\right)\right)\right] = -E\left[J_T^{(\vartheta^*)}\right]$$

so that

$$u\,(0, c - B; \alpha) = -J_0\, e^{-\alpha c}.$$

A method for constructing a candidate family of stochastic processes $\{\, J^{(\vartheta)} \,|\, \vartheta \in \Theta_K \,\}$ satisfying these conditions is examined in the next section.

8.4 Utility indifference hedging and pricing using BSDEs

In order to solve the utility maximisation problem (8.4), we need to identify a family of processes $\left\{ J^{(\vartheta)} \middle| \vartheta \in \Theta_K \right\}$ that satisfies the martingale optimality principle introduced in Theorem 8.5. The problem without the claim (8.3) can be approached using the same method by setting $B = 0$.

8.4.1 *Backward stochastic differential equations*

The main tool in constructing the family $\left\{ J^{(\vartheta)} \middle| \vartheta \in \Theta_K \right\}$ is quadratic backward stochastic differential equations (BSDEs) driven by continuous semimartingales. A key technical problem with BSDEs is assessing whether a unique and sufficiently integrable adapted solution exists.

The Kunita-Watanabe inequality implies that there exist an increasing and bounded process C and a predictable $\mathbb{R}^{n \times n}$-valued 'volatility matrix', denoted σ, such that the matrix $\sigma\sigma'$ is invertible and

$$d\left[M\right] = \sigma\sigma' \, dC.$$

Moreover, assume that there exist a constant $k > 0$ such that

$$\int_0^T \lambda'_t \, d\left[M\right]_t \lambda_t = \int_0^T \left\|\sigma_t \lambda_t\right\|^2 dC_t \le k.$$

Informally, the following dynamics constitute a quadratic BSDE:

$$-dY_t = f\left(t, Z_t\right) dC_t + \frac{\alpha}{2} d\left[L\right]_t - Z'_t \, dM_t - dL_t, \quad t \in [0, T],$$

$$Y_T = \xi. \tag{8.10}$$

The input is a bounded random variable ξ and a random field f, whereas the triplet (Y, Z, L) is the solution.

Definition 8.6. Consider a random variable $\xi \in L^\infty \left(\mathcal{F}_T; P\right)$ and a real-valued random field $f\left(t, z\right) \equiv f\left(t, \omega, z\right)$ such that for all $z \in \mathbb{R}^n$, $f\left(t, z\right)$ is a predictable process. Assume that there exist a constant $\gamma \ge \alpha$ (where α is the risk aversion parameter) and a non-negative predictable process β satisfying $\int_0^T \beta_t \, dC_t \le K$ for a constant $K > 0$ such that

$$\left|f\left(t, z\right)\right| \le \beta_t + \frac{\gamma}{2} \left\|z'\sigma_t\right\|^2$$

and for each $z_1, z_2 \in \mathbb{R}^n$ there exists $C > 0$ such that

$$\left|f\left(t, z_1\right) - f\left(t, z_2\right)\right| \le C \left(\left\|\sigma_s \lambda_s\right\| + \left\|\sigma_s z_1\right\| + \left\|\sigma_s z_2\right\|\right) \left\|\sigma_s \left(z_1 - z_2\right)\right\|.$$

We refer to ξ as the **terminal condition** and f as the **generator**.

Specifying dynamics such as (8.10) is not enough to be able to work with BSDEs as these dynamics are only meaningful when the resulting integrals are well-defined. The next definition specifies a class of solutions which demands not only that the stochastic integrals are well-defined but that they are square-integrable.

Definition 8.7. Let (Y, Z, L) be a triplet of $\mathbb{F}$-adapted processes such that $L \in \mathcal{M}_0^2$, $\left[M^i, L\right] = 0$ for each $1 \leq i \leq n$; Z satisfies

$$E\left[\int_0^T \|\sigma_t Z_t\|^2 \, dC_t\right] < \infty, \quad E\left[\int_0^T |f(t, Z_t)| \, dC_t\right] < \infty;$$

and Y is bounded with

$$Y_t = \xi + \int_t^T f(s, Z_s) \, dC_s - \int_t^T Z_s' \, dM_s + \frac{\alpha}{2}\left([L]_T - [L]_t\right) - (L_T - L_t),$$

$$Y_T = \xi. \tag{8.11}$$

Then the triplet (Y, Z, L) is called the adapted solution to the quadratic **BSDE** (8.10) with terminal condition ξ and generator f.

At this point no trading constraints have been imposed but the square-integrable martingale L describes a part of the random variable ξ that cannot be attained by trading in M even in absence of trading constraints.

Example 8.8. Consider a market driven by a multi-dimensional Brownian motion W with independent components. Let $dM = \sigma \, dW$ for a predictable matrix valued process σ which is full rank and take $C \equiv t$. Define the asset price process as $S = \mathcal{E}(Y)$ where $dY = dM + \lambda' d[M]$. In this market, the PRP implies that $L \equiv 0$ so (8.10) should be taken as

$$-dY_t = f(t, Z_t) \, dt - Z_t' \sigma_t \, dW_t, \quad t \in [0, T], \tag{8.12}$$

$$Y_T = \xi.$$

It is reasonable to propose that to construct a suitable family of processes $\left(J^{(\vartheta)} : \vartheta \in \Theta_K\right)$ which satisfy the martingale optimality principle we should take

$$J^{(\vartheta)} \triangleq \exp\left(-\alpha\left(V^{(\vartheta)} - Y\right)\right) \tag{8.13}$$

where the process Y is the first component of the solution triplet of the quadratic BSDE from Definitions 8.6 and 8.7 with terminal condition $Y_T = B$ and a yet to be identified generator $f(t, z)$. The form of this candidate family enables us to understand the presence of the quadratic variation term $[L]$ in (8.10). Observe that if $\vartheta^* \in \Theta_K$ is such that $J^{(\vartheta^*)}$ is a square integrable $\mathbb{F}$-martingale then it has a GKW decomposition

$$J^{(\vartheta^*)} = \int K'\, dM + \widetilde{L}$$

where $K \in L^2(M)$ and $\widetilde{L} \in \mathcal{M}_0^2$ such that $\left[M^i, \widetilde{L} \right] = 0$ for $1 \leq i \leq n$. This suggests we should attempt to find a random field $g(t, u, k) = g(t, \omega, u, k)$ such that the process U defined via

$$-dU_t = g\left(t, U_t, K_t\right)\, dC_t - K_t'\, dM_t - d\widetilde{L}_t$$

$$U_T = e^{\alpha B}$$

ensures that $J^{(\vartheta)} = U \exp\left(-\alpha V^{(\vartheta)}\right)$ is a super-martingale for all $\vartheta \in \Theta_K$ and a martingale for some $\vartheta \in \Theta_K$. In fact, these dynamics are an exponential transformation of the dynamics defined as in (8.10). This connection can be verified by setting $K = UZ$, $d\widetilde{L} = U\, dL$ and $g(t, u, k) = uf\left(t, k/u\right) - \alpha \left| u \right|^2 \|\sigma\|^2 /2$ and applying the Itô formula to $1/\alpha \log U$. Thus, the appearance of the quadratic variation of $[L]$ in (8.10) is due to the difference between the stochastic and ordinary logarithm. The quadratic variation $[M]$ does not feature explictly in (8.10) as it can be expressed as a dC-integral so may be included in $f(t, z)\, dC$.

8.4.2 *Maximising utility from terminal wealth under trading constraints*

As discussed in the previous section, we aim to specify a generator $f(t, z)$ such that the candidate family of processes $\left(J^{(\vartheta)} : \vartheta \in \Theta_K\right)$ defined via

$$J^{(\vartheta)} \triangleq \exp(-\alpha(V^{(\vartheta)} - Y)) \qquad (8.14)$$

satisfies the martingale optimality principle where the process Y is the first component of the solution of a quadratic BSDE with terminal condition $Y_T = B$. Using the wealth process $V^{(\vartheta)}$ defined in (8.1) and Y defined as in (8.18), each process $J^{(\vartheta)}$ can be decomposed as follows:

$$J^{(\vartheta)} = M^{(\vartheta)} A^{(\vartheta)}$$

where $M^{(\vartheta)}$ is the local martingale

$$M^{(\vartheta)} = J_0^{(\vartheta)} \mathcal{E}\left(-\alpha \int \left(\operatorname{diag}(S)\, \vartheta - Z\right)'\, dM + \alpha L\right), \qquad (8.15)$$

and

$$A^{(\vartheta)} = \exp\left(\int v\left(s, \operatorname{diag}\left(S\right)\vartheta, Z\right) dC\right),$$

where

$$v\left(t, \varphi, z\right) = \frac{\alpha^2}{2} \left\|\sigma_t\left(\varphi - z\right)\right\|^2 - \alpha\left(\sigma_t\varphi\right)'\left(\sigma_t\lambda_t\right) - \alpha f\left(t, z\right).$$

To ensure that for each $\vartheta \in \Theta_K$ the process $A^{(\vartheta)}$ is an increasing process, we require that

$$0 \le \frac{\alpha}{2} \left\|\sigma_t\left(\varphi - z\right)\right\|^2 - \left(\sigma_t\varphi\right)'\left(\sigma_t\lambda_t\right) - f\left(t, z\right) \tag{8.16}$$

$$= \frac{\alpha}{2} \left\|\sigma_t\left(\varphi - \left(z + \frac{\lambda_t}{\alpha}\right)\right)\right\|^2 - \left(\sigma_t z\right)'\left(\sigma_t\lambda_t\right) - \frac{1}{2\alpha}\left\|\sigma_t\lambda_t\right\|^2 - f\left(t, z\right),$$

which is the case when

$$f\left(t, z\right) = \inf_{\vartheta \in \Theta_K} \frac{\alpha}{2} \left\|\sigma_t\left(\operatorname{diag}\left(S_t\right)\vartheta - \left(z + \frac{\lambda_t}{\alpha}\right)\right)\right\|^2 \tag{8.17}$$

$$- \left(\sigma_t z\right)'\left(\sigma_t\lambda_t\right) - \frac{1}{2\alpha}\left\|\sigma_t\lambda_t\right\|^2.$$

This choice of generator does not depend on the strategy and hence the initial value $J_0^{(\vartheta)} = \exp\left(-\alpha\left(c - Y_0\right)\right)$ is independent of the strategy. This argument gives a heuristic idea of how to use the martingale optimality principle to find the maximum expected utility in (8.4); the next theorem provides the required details.

Theorem 8.9. Constrained utility maximisation. *Let the constraint set K be convex and suppose that there exists (Y, Z, L) satisfying the BSDE (8.10) with generator $f\left(t, z\right)$ given in (8.17) and $\xi = B$ so that*

$$Y_t = B + \int_t^T f\left(s, Z_s\right) dC_s - \int_t^T Z_s' dM_s - \int_t^T dL_s + \frac{\alpha}{2}\int_t^T d\left[L\right]_s \tag{8.18}$$

is well-defined. The solution to the portfolio allocation problem (8.4) is

$$u(t, c - B; \alpha) = -\exp\left(-\alpha\left(c - Y_t\right)\right), \tag{8.19}$$

and the unique strategy achieving the supremum in (8.6), denoted $\vartheta^* \in \Theta_K$, is such that

$$\vartheta_t^* \in \arg\min_{\vartheta \in \Theta_K} \left\|\sigma_t\left(\operatorname{diag}\left(S_t\right)\vartheta - \left(Z_t + \frac{\lambda_t}{\alpha}\right)\right)\right\|^2 \tag{8.20}$$

for each $t \in [0, T]$.

Proof. Denote by $(\tau^n)_{n\geq 1}$ a sequence of stopping times such that for each n the process $\left(J^{(\vartheta)}\right)^{\tau^n}$ is a sub-martingale and the process $\left(M^{(\vartheta)}\right)^{\tau^n}$ is a martingale, for each $\vartheta \in \Theta_K$. In particular, for $s \in [t, T]$ and $A \in \mathcal{F}_t$

$$E\left[J_{s\wedge\tau^n}^{(\vartheta)}\mathbb{I}_A\right] \geq E\left[J_{t\wedge\tau^n}^{(\vartheta)}\mathbb{I}_A\right]. \tag{8.21}$$

Moreover, as a consequence of Definition 8.1(*iii*) the family of random variables $\left\{J_{s\wedge\tau^n}^{(\vartheta)}\right\}$ is uniformly integrable since by definition the process Y satisfying (8.18) is bounded. Hence, by letting $n \to \infty$ in (8.21) we achieve that for $s \in [t, T]$ and $A \in \mathcal{F}_t$

$$E\left[J_s^{(\vartheta)}\mathbb{I}_A\right] \geq E\left[J_t^{(\vartheta)}\mathbb{I}_A\right],$$

and consequently for each $\vartheta \in \Theta_K$ the process $J^{(\vartheta)}$ is a sub-martingale. Moreover, taking $\vartheta^* \in \Theta_K$ such that (8.20) holds gives $A^{(\vartheta^*)} \equiv 1$ and

$$J^{(\vartheta^*)} = M^{(\vartheta^*)} = J_0\,\mathcal{E}\left(-\alpha\int\left(\mathrm{diag}\,(S)\,\vartheta^* - Z\right)'\,dM\right).$$

This stochastic exponential will be a martingale if both $\int Z'\,dM$ and $\int \left(\mathrm{diag}\,(S)\,\vartheta^*\right)'\,dM$ are BMO-martingales, see Section 2.5. Take a constant $c > 0$ such that $|Y_t| \leq c$ for all $t \in [0, T]$. Applying the Itô formula to $(Y - c)^2$ and substituting in (8.11) gives for every stopping time $\tau \in [0, T]$

$$(c - B)^2 - (c - Y_\tau)^2 = -2\int_\tau^T (c - Y_t)\, Z_t'\, dM_t - 2\int_\tau^T (c - Y_t)\, dL_t$$

$$+ 2\int_\tau^T (c - Y_t)\, f\,(t, Z_t)\, dC_t$$

$$+ \alpha\int_\tau^T (1 + c - Y_t)\, d\,[L]_t + \int_\tau^T \|\sigma_t Z_t\|^2\, dC_t.$$

By taking conditional expectations and using the integrability imposed in Definition 8.7 we have

$$E\left[\int_\tau^T \|\sigma_t Z_t\|^2\, dC_t + \int_\tau^T (1 + c - Y_t)\, d\,[L]_t\,\middle|\,\mathcal{F}_\tau\right]$$

$$= E\left[(c - B)^2\,\middle|\,\mathcal{F}_\tau\right] - (c - Y_\tau)^2 - E\left[2\int_\tau^T (c - Y_t)\, f\,(t, Z_t)\, dC_t\,\middle|\,\mathcal{F}_\tau\right].$$

From the definition of $f\,(t, z)$ in (8.17) we get

$$-f\,(t, z) \leq (\sigma_t z)'\,(\sigma_t\lambda_t) + \frac{1}{2\alpha}\|\sigma_t\lambda_t\|^2$$

and $(1 + c - Y_t) > 0$, so there exists a constant $c_1 > 0$ such that

$$E\left[\int_\tau^T \|\sigma_t Z_t\|^2 \, dC_t \, \middle| \, \mathcal{F}_\tau\right] \leq c^2 - E\left[2\int_\tau^T (c - Y_t) f(t, Z_t) \, dC_t \, \middle| \, \mathcal{F}_\tau\right]$$

$$\leq c^2 + 2cE\left[\int_\tau^T (\sigma_t Z_t)'(\sigma_t \lambda_t) \, dC_t \, \middle| \, \mathcal{F}_\tau\right]$$

$$+ \frac{c}{\alpha} E\left[\int_\tau^T \|\sigma_t \lambda_t\|^2 \, dC_t \, \middle| \, \mathcal{F}_\tau\right]$$

$$\leq c_1 + \frac{1}{2} E\left[\int_\tau^T \|\sigma_t Z_t\|^2 \, dC_t \, \middle| \, \mathcal{F}_\tau\right]. \qquad (8.22)$$

The final inequality follows since

$$(\sigma_t Z_t)'(\sigma_t \lambda_t) \leq \frac{1}{4c} \|\sigma_t Z_t\|^2 + 4c \|\sigma_t \lambda_t\|^2$$

and by assumption

$$E\left[\int_\tau^T \|\sigma_t \lambda_t\|^2 \, dC_t \, \middle| \, \mathcal{F}_\tau\right] \leq E\left[\int_\tau^T \lambda_t' \, d[M]_t \, \lambda_t \, \middle| \, \mathcal{F}_\tau\right] \leq k. \qquad (8.23)$$

Recalling that $d[M] = \sigma\sigma' \, dC$, (8.22) shows that there exists a constant $c_2 > 0$, independent of τ, such that

$$E\left[\int_\tau^T Z_t' \, d[M]_t \, Z_t \, \middle| \, \mathcal{F}_\tau\right] \leq c_2$$

which implies that $\int Z' \, dM$ is a BMO-martingale. To show that $\int (\mathrm{diag}\,(S)\,\vartheta^*)' \, dM \in BMO$ as well, observe that the triangle inequality gives

$$\|\sigma_s \mathrm{diag}\,(S_s)\,\vartheta_s^*\| \leq \left\|\sigma_s\left(\mathrm{diag}\,(S_s)\,\vartheta_s^* - \left(Z_s + \frac{\lambda_s}{\alpha}\right)\right)\right\| + \left\|\sigma_s\left(Z_s + \frac{\lambda_s}{\alpha}\right)\right\|.$$

However, since $0 \in K$,

$$\left\|\sigma_s\left(\mathrm{diag}(S_s)\vartheta_s^* - \left(Z_s + \frac{\lambda_s}{\alpha}\right)\right)\right\| = \inf_{\vartheta \in \Theta_K} \left\|\sigma_s\left(\mathrm{diag}(S_s)\vartheta_s - \left(Z_s + \frac{\lambda_s}{\alpha}\right)\right)\right\|$$

$$\leq \left\|\sigma_s\left(Z_s + \frac{\lambda_s}{\alpha}\right)\right\|,$$

hence

$$\|\sigma_s \mathrm{diag}\,(S_s)\,\vartheta_s^*\| \leq 2\left\|\sigma_s\left(Z_s + \frac{\lambda_s}{\alpha}\right)\right\| \leq \frac{2}{\alpha}\|\sigma_s \lambda_s\| + 2\|\sigma_s Z_s\|.$$

Subsequently, using that $(a + b)^2 \leq 2 \left(a^2 + b^2 \right)$ and (8.23)

$$E \left[\int_\tau^T \| \sigma_s \mathrm{diag} \left(S_s \right) \vartheta_s^* \|^2 \, dC_s \,\middle|\, \mathcal{F}_\tau \right]$$

$$\leq 4E \left[\int_\tau^T \left(\frac{1}{\alpha} \| \sigma_s \lambda_s \| + \| \sigma_s Z_s \| \right)^2 \, dC_s \,\middle|\, \mathcal{F}_\tau \right]$$

$$\leq 8E \left[\int_\tau^T \left(\frac{1}{\alpha^2} \| \sigma_s \lambda_s \|^2 + \| \sigma_s Z_s \|^2 \right) \, dC_s \,\middle|\, \mathcal{F}_\tau \right]$$

$$\leq c + 8E \left[\int_\tau^T Z_s' \, d \left[M \right]_s Z_s \,\middle|\, \mathcal{F}_\tau \right].$$

We may conclude that $\int (\mathrm{diag}(S)\vartheta^*)' \, dM$ is a BMO-martingale since it has been shown that $\int Z' \, dM$ is a BMO-martingale. Due to Theorem 2.46, these estimates show that

$$M^{(\vartheta^*)} = J_0 \, \mathcal{E} \left(-\alpha \int (\mathrm{diag} \left(S \right) \vartheta^* - Z)' \, dM \right)$$

is a martingale and that the strategy ϑ^* defined in (8.20) is admissible in the sense of Definition 8.1. Moreover, by applying the martingale optimality principle as outlined in Theorem 8.5, we may conclude that $\vartheta^* \in \Theta_K$ is the strategy solving the portfolio allocation problem (8.4). This strategy is unique due to the assumption that K is convex, since this implies that at each $t \in [0, T]$ (8.20) contains only one element. $\square$

It is important to note that the method discussed so far can also be used to solve the portfolio allocation problem without the claim (8.3) by setting $B = 0$. In this case we should search for a solution $(\widetilde{Y}, \widetilde{Z}, \widetilde{L})$ to a BSDE with terminal condition $\xi = 0$ and generator defined as in (8.17), that is

$$\widetilde{Y}_t = \int_t^T f(s, \widetilde{Z}_s) \, dC_s - \int_t^T \widetilde{Z}_s' \, dM_s - \int_t^T d\widetilde{L}_s + \frac{\alpha}{2} \int_t^T d \left[\widetilde{L} \right]_s . \quad (8.24)$$

Then the trading strategy $\vartheta^0 \in \Theta_K$ which achieves the supremum in (8.3) satisfies

$$\vartheta_t^0 \in \arg \min_{\vartheta \in \Theta_K} \left\| \sigma_t \left(\mathrm{diag} \left(S_t \right) \vartheta - \left(\widetilde{Z}_t + \frac{\lambda_t}{\alpha} \right) \right) \right\|^2 .$$

Furthermore, it follows from its definition that the utility indifference price process introduced in Definition 8.4 can be expressed as

$$\pi = Y - \widetilde{Y}.$$

The adjustment in strategy necessary to hedge B can be constructed from the strategies ϑ^* and ϑ^0 using Definition 8.3,

$$\vartheta^B = \left(\vartheta^* - \vartheta^0 \right).$$

8.5   Examples in Brownian markets

In this section we assume that the filtration $\mathbb{F}$ is the completion of the natural filtration of a two-dimensional Brownian motion $W = \left(W^1, W^2 \right)'$ such that $\left[W^1, W^2 \right] = \rho t$ with correlation process $|\rho| \in (0, 1]$. The financial market is assumed to consist of two assets:

$$dS = S \left(\mu \, dt + \sigma \, dW^1 \right),$$
$$dP = P \left(\beta \, dt + \gamma \, dW^2 \right).$$

The parameters μ/σ, β/γ are bounded $\mathbb{F}$-predictable processes. Furthermore, we assume that σ and γ are bounded away from zero which in particular implies that, in the absence of trading constraints, this market is complete. To prevent arbitrage opportunities we require that either $|\rho| < 1$ or that $\mu = \beta\sigma/\gamma$ on the set $\{ t \, | \rho_t = 1 \}$. The vector of price processes $\widetilde{S} = (S, P)'$ can be expressed as $\widetilde{S} = \mathcal{E}(Y)$ where

$$dY = dM + \lambda' d\,[M]$$

and

$$dY = \begin{pmatrix} dS/S \\ dP/P \end{pmatrix}, \quad M = \sigma W, \quad \lambda' = \left(\mu/\sigma^2, \beta/\gamma^2 \right),$$

with $W = \left(W^1, W^2 \right)'$. Furthermore, with a slight abuse of notation, we denote by σ the 'volatility matrix'

$$\sigma := \begin{pmatrix} \sigma & 0 \\ 0 & \gamma \end{pmatrix}.$$

The orthonormal basis of risk factors is a standard two-dimensional Brownian motion $N = \left(W^1, W^\perp \right)'$ with independent components so that $\left[W^1, W^\perp \right] = 0$. The correlated Brownian motion W^2 can be expressed as $W^2 = \rho W^1 + \sqrt{1 - \rho^2} W^\perp$.

In the first subsection it will be shown that when no trading constraints are present, the method presented in this chapter is consistent with standard risk neutral valuation in complete markets. The second part of this section examines the case of 'basis risk' which occurs in markets with untraded underlying assets such as weather options or certain types of commodity markets.

8.5.1 *Complete markets*

When $K = \mathbb{R}^2$ there are no constraints on the trading strategies and under the assumptions outlined above the market is complete. In this case we can illustrate that the utility indifference price and hedge coincide with the risk neutral price and hedge.

It was shown in Theorem 8.9 that if a bounded claim B has been sold then

$$u(c - B; \alpha) = -\exp\left(-\alpha\left(c - Y_0\right)\right)$$

where Y_0 is the unique initial value of the BSDE

$$Y_t = B + \int_t^T f\left(s, Z_s\right) ds - \int_t^T Z_s' \sigma_s \, dW_s. \tag{8.25}$$

In this case $Z = \left(Z^1, Z^2\right)'$ and the generator (8.17) reads

$$f\left(t, z\right) := -\left(\sigma_t z\right)'\left(\sigma_t \lambda_t\right) - \frac{1}{2\alpha} \left\|\sigma_t \lambda_t\right\|^2 \tag{8.26}$$

because for each $t \in [0, T]$

$$\inf_{\vartheta \in \mathbb{R}^2} \frac{\alpha}{2} \left\| \sigma_t \left(\operatorname{diag}\left(S_t\right) \vartheta - \left(Z_t + \frac{\lambda}{\alpha}\right)\right) \right\|^2 = 0.$$

Moreover, the maximum expected utility is achieved by the trading strategy ϑ^* such that

$$\left(\operatorname{diag}\left(S\right) \vartheta^*\right)' = \left(S\vartheta^{*,S}, \, P\vartheta^{*,P}\right) = \left(Z^1 + \mu/\alpha\sigma^2, \, Z^2 + \beta/\alpha\gamma^2\right).$$

Similarly, setting $B = 0$ in Theorem 8.9 gives

$$u(c; \alpha) = -\exp(-\alpha(c - \widetilde{Y}_0)),$$

where $\widetilde{Y}_0$ is the unique initial value of the BSDE

$$\widetilde{Y}_t = \int_t^T f(s, \widetilde{Z}_s) ds - \int_t^T \widetilde{Z}_s' \sigma_s \, dW_s. \tag{8.27}$$

Here $\widetilde{Z} = (\widetilde{Z}^1, \widetilde{Z}^2)'$ and the generator is as defined in (8.26). The maximum expected utility is achieved by the trading strategy ϑ^0 such that

$$\left(\text{diag}\,(S)\,\vartheta^0\right)' = \left(S\vartheta^{0,S},\,P\vartheta^{0,P}\right) = (\,\widetilde{Z}^1 + \mu/\alpha\sigma^2,\,\widetilde{Z}^2 + \beta/\alpha\gamma^2\,).$$

From Definition 8.3 we have that the utility indifference hedging strategy for the claim B, denoted ϑ^B, satisfies

$$\left(\text{diag}\,(S)\,\vartheta^B\right)' = \left(S\vartheta^{B,S},\,P\vartheta^{B,P}\right) = (\,Z^1 - \widetilde{Z}^1,\,Z^2 - \widetilde{Z}^2\,).$$

More insight can be gained into the utility indifference price and hedge in this unconstrained case because BSDEs with a linear generator can be explicitly solved. The solution to the BSDEs (8.25) and (8.27) can be expressed as a conditional expectation. This representation illustrates an intrinsic link between BSDEs and Feynman-Kac-type representations. Commonly, BSDEs with general generators are used as a non-linear generalisation of the Feynman-Kac representation, see Yong & Zhou (1999).

Proposition 8.10. *Explicit solution to linear BSDEs.* *Consider a pair of adapted processes Y and $Z = \left(Z^1, Z^2\right)'$ such that Y is bounded and*

$$E\left[\int_0^T \|\sigma Z_t\|^2\,dt\right] < \infty;$$

and a terminal condition $\xi \in L^2\left(P; \mathcal{F}_T\right)$ as well as a linear generator

$$f\,(t,k) := a + (\sigma z)'\,b$$

where both a and $b = \left(b^1, b^2\right)'$ are bounded. Then (Y, Z) can be represented as

$$-dY = f\,(t,Z)\,dt - Z'\sigma\,dW,$$
$$Y_T = \xi,$$

if and only if the process Y satisfies

$$H_t Y_t = E\left[H_T \xi + \int_t^T H_s a_s\,ds\,\middle|\,\mathcal{F}_t\right]$$

where H is a martingale determined by the dynamics

$$dH = Hb'\,dW,\quad H_0 = 1.$$

Proof. The integration by parts formula implies that

$$d\,(HY) = -Ha\,dt - H\,(Yb' - Z'\sigma)\,dW$$

and hence

$$HY + \int Ha\,ds = Y_0 - \int H\,(Yb' - Z'\sigma)\,dW.$$

The integrability conditions in the definition of a solution to a BSDE ensure that this stochastic integral is a martingale. Hence, $Z \in L^2\,(W)$ is the process such that the martingale

$$M_t = E\left[H_T\xi + \int_0^T H_s a_s\,ds\,\middle|\,\mathcal{F}_t\right]$$

can be represented as

$$M_t = Y_0 + \int_0^t H_s\,(Y_s b'_s - Z'_s\sigma_s)\,dW_s.$$

In this case, since $Y_T = \xi$, we have

$$H_t Y_t + \int_0^t H_s a_s\,ds = E\left[H_T\xi + \int_0^T H_s a_s\,ds\,\middle|\,\mathcal{F}_t\right]$$

from which the result follows. $\qquad\square$

Returning to our example, taking $a = -\|\sigma\lambda\|^2/2\alpha$ and $b = -\sigma\lambda$ in Proposition 8.10 yields

$$Y_t = E_Q\left[B - \frac{1}{2\alpha}\int_t^T \|\sigma_s\lambda_s\|^2\,ds\,\middle|\,\mathcal{F}_t\right]$$

as well as

$$\widetilde{Y}_t = E_Q\left[-\frac{1}{2\alpha}\int_t^T \|\sigma_s\lambda_s\|^2\,ds\,\middle|\,\mathcal{F}_t\right]$$

where the measure $Q \sim P$ is the unique equivalent martingale measure defined via the density process $H = \mathcal{E}\left(-\int \lambda\,dM\right)$. We may conclude that the utility indifference price process is

$$\pi_t = Y_t - \widetilde{Y}_t = E_Q\,[B\,|\mathcal{F}_t]\,, \qquad (8.28)$$

so that the utility indifference price coincides with the risk-neutral value. Moreover, under the measure Q the process $B = (B^1, B^2)'$ with

$$B^1 := W^1 + \int \frac{\mu}{\sigma}\,dt, \quad B^2 := W^2 + \int \frac{\beta}{\gamma}\,dt$$

is a two-dimensional correlated Brownian motion and

$$dY := \begin{pmatrix} dS/S \\ dP/P \end{pmatrix} = \begin{pmatrix} \sigma B^1 \\ \gamma B^2 \end{pmatrix}.$$

Substituting the BSDEs (8.25) and (8.27) into (8.28) gives

$$\pi_t = E_Q\left[B| \mathcal{F}_t\right]$$

$$= B + \int_t^T (\sigma_s(\tilde{Z}_s - Z_s))'(\sigma_s \lambda_s)\, ds + \int_t^T (\tilde{Z}_s - Z_s)' \sigma_s\, dW_s$$

$$= B - \int_t^T (Z_s - \tilde{Z}_s)' \sigma_s\, dB_s = B - \int_t^T \vartheta_s^{B,S}\, dS_s - \int_t^T \vartheta_s^{B,P}\, dP_s.$$

Subsequently, the utility indifference hedge also coincides with the risk-neutral hedge. Specifically, if the claim is of the form $B = h(S_T)$ then $\vartheta^B = (\varphi, 0)'$ where $\varphi \in L^2(B^1)$ is the integrand such that $h(S_T)$ can be represented as

$$h(S_T) = E_Q\left[h(S_T)\right] + \int_0^T \varphi_s\, dS_s.$$

8.5.2   *Basis risk*

Consider a firm which produces a commodity which has price process P and this output is correlated with an exchange traded asset denoted S. The firm has sold a European option on its own output to be delivered at time T with payoff $h(P_T)$.

There are three possible scenarios:

(1) If the firm can hold an inventory of its output and the market is complete, then there is no need to trade in S in order to hedge the claim $h(P_T)$. The company would use the strategy $\vartheta^B = (0, \varphi)'$ where $\varphi \in L^2(W^2)$ is such that

$$h(P_T) = E_Q\left[h(P_T)\right] + \int_0^T \varphi_s\, dP_s. \qquad (8.29)$$

(2) When $\rho \equiv 1$ the two assets are perfect substitutes and the market is complete. To ensure that an equivalent martingale measure exists we require that $\mu = \beta\sigma/\gamma$, so that

$$\frac{dP}{P} = \frac{\sigma}{\gamma}\frac{dS}{S}.$$

In this case, the firm can perfectly replicate the claim $h(P_T)$ using the strategy $\vartheta^B = (\varphi\sigma/\gamma, 0)'$ where $\varphi \in L^2(W^1)$ is the process defined via (8.29) since

$$h(P_T) = E_Q[h(P_T)] + \int_0^T \frac{\sigma}{\gamma}\varphi_s \, dS_s.$$

(3) In the case that $|\rho| \in (0,1)$ and the company is unable to keep an inventory then it is less clear how to hedge the option by trading only in the traded asset. In this case, not being able to hold the asset P to hedge the claim is a trading constraint which makes the market incomplete. The strategy space is

$$K = \left\{ (x,y) \in \mathbb{R}^2 \,\middle|\, y = 0 \right\}.$$

As P is untradeable, it is not required to be a martingale under the equivalent martingale measure. The drift of asset P under Q can be arbitrarily chosen so there is no unique equivalent martingale measure.

In the rest of this example we focus on contructing an appropriate hedge in the third case by applying Theorem 8.9. In this example, the BSDE used in Theorem 8.9 depends on the forward SDE of the untraded asset P, so the utility maximisation problem is based on the following coupled system:

$$\begin{aligned} dP &= P\left(\beta \, dt + \gamma \, dW^2\right) \\ -dY &= f(t,Z) \, dt - Z'\sigma \, dW \\ P_0 &= p, \quad Y_T = h(P_T), \end{aligned} \tag{8.30}$$

where $Z = (Z^1, Z^2)'$. This type of system is referred to as a forward backward stochastic differential equation (FBSDE). The addition of forward dynamics complicate showing that a unique solution exists as the problem is now a generalisation of a two-point boundary problem. As in Theorem 8.9, the generator should be taken to be

$$\begin{aligned} f(t,z) &:= \inf_{\vartheta \in \Theta_K} \frac{\alpha}{2} \left\| \sigma_t \left(\text{diag}(S_t)\vartheta - \left(z + \frac{\lambda_t}{\alpha}\right) \right) \right\|^2 \\ &\quad - (\sigma_t z)'(\sigma_t \lambda_t) - \frac{1}{2\alpha}\|\sigma_t \lambda_t\|^2 \\ &= \frac{\alpha}{2}\left(\gamma z_2 + \frac{\beta}{\alpha\gamma}\right)^2 - \mu z_1 - \beta z_2 - \frac{1}{2\alpha}\frac{\mu^2}{\sigma^2} - \frac{1}{2\alpha}\frac{\beta^2}{\gamma^2} \\ &= \frac{\alpha}{2}\gamma^2 z_2^2 - \mu z_1 - \frac{1}{2\alpha}\frac{\mu^2}{\sigma^2} \end{aligned} \tag{8.31}$$

and the corresponding trading strategy which maximises utility from terminal wealth satisfies

$$(\operatorname{diag}(S)\,\vartheta^*)' = (S\vartheta^{*,S},0) = (Z^1 + \mu/\alpha\sigma^2,0). \qquad (8.32)$$

On the other hand, suppose that the firm had not sold the claim $F(P_T)$. In this case the FBSDE (8.30) reads

$$dP = P\left(\beta\,dt + \gamma\,dW^2\right)$$

$$-d\widetilde{Y} = f(t,\widetilde{Z})\,dt - \widetilde{Z}'\sigma\,dW$$

$$P_0 = p, \quad \widetilde{Y}_T = 0,$$

where $\widetilde{Z} = (\widetilde{Z}^1,\widetilde{Z}^2)'$ and the generator is as in (8.31). Applying Theorem 8.9, the trading strategy which maximises utility from terminal wealth satisfies

$$\left(\operatorname{diag}(S)\,\vartheta^0\right)' = (S\vartheta^{0,S},0) = (\widetilde{Z}^1 + \mu/\alpha^2,0),$$

and the utility indifference hedge satisfies

$$\left(\operatorname{diag}(S)\,\vartheta^B\right)' = (S\vartheta^{B,S},0) = (Z^1 - \widetilde{Z}^1,0).$$

Due to the first term in the generator (8.31), it is not possible to apply Proposition 8.10 to find an explicit solution to the quadratic BSDE. Instead, consider the process $U = \exp(\alpha Y)$ which has the following linear dynamics

$$dU = \left(\alpha\mu K^1 + \frac{1}{2}\frac{\mu^2}{\sigma^2}U\right)dt + \alpha\sigma K^1\,dW^1 + \alpha\gamma K^2\,dW^2$$

with $K^i = UZ^i$ for $i = 1,2$ and terminal condition $U_T = \exp(\alpha h(P_T))$. To handle this BSDE, we switch to an orthonormal basis by setting

$$W^2 = \rho W^1 + \sqrt{1-\rho^2}W^\perp$$

with $[W^1, W^\perp] = 0$ and change to a measure Q with density

$$\frac{dQ}{dP} = \mathcal{E}\left(-\int\frac{\mu}{\sigma}\,dW^1\right)\mathcal{E}\left(\int\frac{\rho\mu}{\sigma\sqrt{1-\rho^2}}\,dW^\perp\right).$$

The processes

$$B^1 = W^1 + \frac{\mu}{\sigma}t, \quad B^\perp = W^\perp - \frac{\rho\mu}{\sigma\sqrt{1-\rho^2}}t$$

are Q-Brownian motions and the Q-dynamics of U are

$$dU = \frac{1}{2}\frac{\mu^2}{\sigma^2}U\,dt + \alpha\left(\sigma K^1 + \gamma\rho K^2\right)dB^1 + \alpha\gamma\sqrt{1-\rho^2}K^2\,dB^\perp \qquad (8.33)$$

with terminal condition $U_T = \exp\left(\alpha h\left(P_T\right)\right)$. Applying integration by parts to $U \exp(-\frac{1}{2}\int \frac{\mu^2}{\sigma^2}\, dt)$, integrating and taking conditional expectations we obtain

$$U_t = E_Q\left[\exp\left(\alpha h\left(P_T\right) - \frac{1}{2}\int_t^T \frac{\mu^2}{\sigma^2}\, du\right)\middle|\mathcal{F}_t\right].$$

Similarly, by setting $\widetilde{U} = \exp(\alpha\widetilde{Y})$ we have

$$\widetilde{U}_t = E_Q\left[\exp\left(-\frac{1}{2}\int_t^T \frac{\mu^2}{\sigma^2}\, du\right)\middle|\mathcal{F}_t\right],$$

hence the utility indifference price process is

$$\pi = Y - \widetilde{Y} = \frac{1}{\alpha}(\log U - \log \widetilde{U}).$$

Take a bounded random variable X. Then e^X has the following representation under Q:

$$e^X = E_Q\left[e^X\right] + \int_0^T \varphi_t^1\, dB_t^1 + \int_0^T \varphi_t^\perp\, dB_t^\perp$$

for some $\varphi^1 \in L^2\left(B^1\right)$ and $\varphi^\perp \in L^2\left(B^\perp\right)$. Denote by H the process such that $H_t = E_Q\left[e^X\middle|\mathcal{F}_t\right]$. Applying the Itô formula to $\log H$ we can see that e^X may be represented as

$$e^X = E_Q\left[e^X\right]\mathcal{E}\left(\int \frac{\varphi^1}{H}\, dB^1\right)\mathcal{E}\left(\int \frac{\varphi^\perp}{H}\, dB^\perp\right). \tag{8.34}$$

On the other hand, taking

$$X := \alpha h\left(P_T\right) - \frac{1}{2}\int_0^T \frac{\mu^2}{\sigma^2}\, ds \tag{8.35}$$

and applying the Itô formula to $\log U$ using the dynamics given in (8.33) gives

$$e^X = E_Q\left[e^X\right]\mathcal{E}\left(\int \left(\sigma Z^1 + \gamma\rho Z^2\right) dB^1\right)\mathcal{E}\left(\int \gamma\sqrt{1-\rho^2}Z^2\, dB^\perp\right). \tag{8.36}$$

Comparing (8.34) and (8.36) we obtain

$$Z_t^1 = \frac{\varphi_t^1 - \frac{\rho}{\gamma\sqrt{1-\rho^2}}\varphi_t^\perp}{E_Q\left[X\middle|\mathcal{F}_t\right]}, \quad Z_t^2 = \frac{1}{\gamma\sqrt{1-\rho^2}}\frac{\varphi_t^\perp}{E_Q\left[X\middle|\mathcal{F}_t\right]}$$

where $\varphi^1 \in L^2\left(B^1\right), \varphi^\perp \in L^2\left(B^\perp\right)$ are such that

$$X = E_Q\left[X\right] + \int_0^T \varphi_t^1\, dB_t^1 + \int_0^T \varphi_t^\perp\, dB_t^\perp.$$

Here X is the random variable defined in (8.35). A directly similar argument can be used to show that

$$\tilde{Z}_t^1 = \frac{\psi_t^1 - \frac{\rho}{\gamma\sqrt{1-\rho^2}}\psi_t^\perp}{E_Q\left[-\frac{1}{2}\int_0^T \frac{\mu^2}{\sigma^2}\,ds\,\middle|\,\mathcal{F}_t\right]}, \qquad \tilde{Z}_t^2 = \frac{1}{\gamma\sqrt{1-\rho^2}}\frac{\psi_t^\perp}{E_Q\left[-\frac{1}{2}\int_0^T \frac{\mu^2}{\sigma^2}\,ds\,\middle|\,\mathcal{F}_t\right]}.$$

Here $\psi^1 \in L^2\left(B^1\right), \psi^\perp \in L^2\left(B^\perp\right)$ are such that

$$-\frac{1}{2}\int_0^T \frac{\mu^2}{\sigma^2}\,ds = E_Q\left[-\frac{1}{2}\int_0^T \frac{\mu^2}{\sigma^2}\,ds\right] + \int_0^T \psi_t^1\,dB_t^1 + \int_0^T \psi_t^\perp\,dB_t^\perp.$$

Hence, the utility indifference hedge is $\left(\vartheta^B\right)' = \left(\vartheta^{B,S}, 0\right)$ with

$$\vartheta_t^{B,S} = \frac{\varphi_t^1 - \frac{\rho}{\gamma\sqrt{1-\rho^2}}\varphi_t^\perp}{E_Q\left[\alpha h\left(P_T\right) - \frac{1}{2}\int_0^T \frac{\mu^2}{\sigma^2}\,ds\,\middle|\,\mathcal{F}_t\right]} - \frac{\psi_t^1 - \frac{\rho}{\gamma\sqrt{1-\rho^2}}\psi_t^\perp}{E_Q\left[-\frac{1}{2}\int_0^T \frac{\mu^2}{\sigma^2}\,ds\,\middle|\,\mathcal{F}_t\right]}.$$

8.6 Connection to the minimal entropy measure in the unconstrained case

This section links the material in this chapter with the dual approach taken in the previous chapter. Take $K = \mathbb{R}$, so there are no trading constraints, and for ease of exposition assume that the price process $S = \mathcal{E}\left(Y\right)$ is one-dimensional where $dY = dM + \lambda\,d\left[M\right]$ and

$$\int_0^T \lambda_t^2\,d\left[M\right]_t = \int_0^T \sigma_t^2\lambda_t^2\,dC_t \leq k. \tag{8.37}$$

In Chapter 7, the strategy space was taken to be

$$\Theta = \left\{\vartheta \in L\left(S\right)\,\middle|\,\int \vartheta\,dS \text{ is a } Q\text{-martingale for all } Q \in \mathcal{M}^f\right\}. \tag{8.38}$$

In particular, the trading strategy shown in Chapter 7 to relate to the density of the minimal entropy measure is not necessarily square-integrable. In contrast, in this chapter a trading strategy $\vartheta \in L^2\left(M\right)$ is admissible if the family of random variables

$$\left\{\exp\left(-\alpha V_\tau^\vartheta\right)\,\middle|\,\tau \leq T \text{ is a stopping time}\right\}$$

is uniformly integrable.

Furthermore, we adopt the standing assumption that $\mathcal{M}^f \cap \mathcal{M}^e \neq \emptyset$ and introduce a dual minimisation problem

$$v(t, -B; \alpha) = \text{ess} \inf_{Q \in \mathcal{M}^f \cap \mathcal{M}^e} E_Q \left[\frac{1}{\alpha} \log \left(\frac{Z_T^Q}{Z_t^Q} \right) - B \middle| \mathcal{F}_t \right] \qquad (8.39)$$

where B is the bounded claim in the utility maximisation problem (8.4) and Z^Q is the density process of Q with respect to P.

In particular, when $B = 0$ and $t = 0$ the infimum in (8.39) is achieved by the minimal entropy martingale measure Q^0, and we have

$$v(0, 0; \alpha) = \frac{1}{\alpha} H(Q^0, P).$$

It was shown in Theorem 7.22 that if $c = 0$, the value of the utility maximisation problem (7.12) is

$$-u(0, 0; \alpha) = \inf_{\vartheta \in \Theta} E \left[\exp \left(-\alpha \int_0^T \vartheta_u \, dS_u \right) \right]$$

$$= \exp \left(-H(Q^0, P) \right) = \exp \left(-\alpha v(0, 0; \alpha) \right),$$

and that the infimum is achieved by the strategy $-\frac{1}{\alpha} \vartheta^0 \in \Theta$. Theorem 7.14 identifies the process $\vartheta^0 \in L(S)$ such that $\int \vartheta^0 \, dS$ is a Q^0-martingale as the integrand such that the density of the entropy measure may be expressed as

$$\frac{dQ^0}{dP} = \exp \left(c_0 + \int_0^T \vartheta_t^0 \, dS_t \right) \qquad (8.40)$$

where c_0 is a constant. This representation only holds at maturity and in general the density process Z^0 of the entropy measure cannot be represented as such an ordinary exponential at intermediary times.

On the other hand, when a bounded claim B has been sold, the infimum of the value of (8.39) at $t = 0$ is

$$v(0, -B; \alpha) = \frac{1}{\alpha} H(Q^\alpha, P^\alpha)$$

where $dP^\alpha / dP = c e^{\alpha B}$ for a suitable normalising constant c and Q^α is the entropy measure relative to P^α. Furthermore, it was shown in Theorem 7.25 that the value of the utility maximisation problem is

$$-u(0, -B; \alpha) = \inf_{\vartheta \in \Theta} E \left[\exp \left(\alpha \left(B - \int_0^T \vartheta_u \, dS_u \right) \right) \right]$$

$$= \exp \left(-H(Q^\alpha, P^\alpha) \right) = \exp \left(-\alpha v(0, -B; \alpha) \right)$$

and the infimum is achieved by the strategy $-\frac{1}{\alpha}\vartheta^\alpha \in \Theta$. The predictable process ϑ^α is the process identified in Theorem 7.14 such that the density of Q^α may be expressed as

$$\frac{dQ^\alpha}{dP^\alpha} = \exp\left(c_\alpha + \int_0^T \vartheta_t^\alpha \, dS_t\right)$$

for a constant c_α and $\vartheta^\alpha \in L(S)$ such that $\int \vartheta^\alpha \, dS$ is a Q^α-martingale.

The next proposition, proved in Mania & Schweizer (2005), generalises the duality principle outlined in Theorem 7.25 to the dynamic framework presented here.

Proposition 8.11. *Duality principle.* *For fixed $\alpha > 0$ the utility max-imisation problem introduced in (8.6),*

$$-u\left(t, -B; \alpha\right) = \operatorname*{ess\ inf}_{\vartheta \in \Theta} E\left[\exp\left(\alpha\left(B - \int_0^T \vartheta_u \, dS_u\right)\right)\Bigg| \mathcal{F}_t\right]$$

and the dual problem (8.39) are related by

$$u\left(t, -B; \alpha\right) = -\exp\left(-\alpha v\left(t, -B; \alpha\right)\right). \tag{8.41}$$

Subsequently, the utility indifference price process for the claim B, in-troduced in Definition 8.4, can be expressed as

$$\pi_t = \frac{1}{\alpha}\log\left(\frac{u\left(t, -B; \alpha\right)}{u\left(t, 0; \alpha\right)}\right) = v\left(t, 0; \alpha\right) - v\left(t, -B; \alpha\right).$$

The next proposition reformulates the utility indifference price as a sin-gle optimisation problem under the entropy measure.

Proposition 8.12. *Primal problem under Q^0.* *The exponential utility price process π may be written as*

$$\exp\left(\alpha\pi_t\right) = \operatorname*{ess\ inf}_{\vartheta \in \Theta} E_{Q^0}\left[\exp\left(\alpha\left(B - \int_t^T \vartheta_u \, dS_u\right)\right)\Bigg| \mathcal{F}_t\right]. \tag{8.42}$$

Proof. From (8.39) and (8.40) we have that

$$\alpha v\left(t, 0; \alpha\right) = E_{Q^0}\left[\log\left(\frac{Z_T^0}{Z_t^0}\right)\Bigg| \mathcal{F}_t\right] = c_0 + \int_0^t \vartheta_u^0 \, dS_u - \log Z_t^0 \tag{8.43}$$

where Z^0 is the density process of Q^0. Define a process $\overline{Z}$ via

$$\overline{Z} = \exp\left(c_0 + \int \vartheta^0\, dS\right);$$

it is such that $\overline{Z}_T = Z^0_T$ but, in general, $\overline{Z}_t \neq Z^0_t$ for $t < T$. We may now rewrite $u(t,0;\alpha)$ using Proposition 8.11 and (8.43) as

$$u(t,0;\alpha) = -\exp(-\alpha v(t,0;\alpha)) = -\frac{Z^0_t}{\overline{Z}_t}. \tag{8.44}$$

Thus, it follows from the definition of the utility indifference price process that for $t \in [0,T]$

$$\exp(\alpha\pi_t) = \frac{u(t,-B;\alpha)}{u(t,0;\alpha)} = u(t,-B;\alpha)\frac{\overline{Z}_t}{Z^0_t}$$

$$= \operatorname*{ess\,inf}_{\vartheta\in\Theta} E\left[\exp\left(\alpha\left(B - \int_t^T \vartheta_u\, dS_u\right)\right)\frac{\overline{Z}_t}{Z^0_t}\frac{Z^0_T}{\overline{Z}_T}\Big|\mathcal{F}_t\right]$$

$$= \operatorname*{ess\,inf}_{\vartheta\in\Theta} E_{Q^0}\left[\exp\left(\alpha\left(B - \int_t^T \left(\vartheta_u + \frac{\vartheta^0_u}{\alpha}\right) dS_u\right)\right)\Big|\mathcal{F}_t\right]$$

$$= \operatorname*{ess\,inf}_{\psi\in\Theta} E_{Q^0}\left[\exp\left(\alpha\left(B - \int_t^T \psi_u\, dS_u\right)\right)\Big|\mathcal{F}_t\right].$$

The final equality holds because $\psi = \vartheta + \vartheta^0/\alpha \in \Theta$ for each $\vartheta \in \Theta$. □

For Proposition 8.12 to hold it is vital that ϑ^0, the strategy defined in (8.40), is in the set of admissible strategies Θ defined in (8.38). In general, ϑ^0 may not be in the constrained strategy set Θ_K introduced in Definition 8.1 for two reasons: (i) it need not be that $\vartheta^0 \in L^2(M)$ and (ii) the strategy space K can be an arbitrary convex set containing the origin, so it may occur that $\exists t \geq 0$ such that $\vartheta^0_t \notin K$.

Recall that in the unconstrained case the portfolio allocation problem (8.3) is shown in Theorem 8.9 to be related to the solution of a BSDE, denoted $(\widetilde{Y}, \widetilde{Z}, \widetilde{L})$, with terminal condition $\xi = 0$ and generator

$$f(t,z) = -z\lambda_t\sigma_t^2 - \frac{1}{2\alpha}\sigma_t^2\lambda_t^2.$$

The process λ is such that the stochastic logarithm of the price process has the canonical decomposition $dS/S = dY = dM + \lambda\, d[M]$. Moreover,

$d\,[M] = \sigma^2\,dC$, so examining the dynamics of $\widetilde{Y}$ from (8.24) yields

$$\widetilde{Y} = \widetilde{Y}_0 - \int f(t,\widetilde{Z})\,dC + \int \widetilde{Z}\,dM + \widetilde{L} - \frac{\alpha}{2}[\widetilde{L}]$$

$$= \widetilde{Y}_0 + \int \widetilde{Z}\lambda\,d\,[M] + \int \frac{1}{2\alpha}\lambda^2\,d\,[M] + \int \widetilde{Z}\,dM + \widetilde{L} - \frac{\alpha}{2}[\widetilde{L}]$$

$$= \widetilde{Y}_0 + \int \frac{1}{2\alpha}\lambda^2\,d\,[M] + \int \widetilde{Z}\,\frac{dS}{S} + \widetilde{L} - \frac{\alpha}{2}[\widetilde{L}] \tag{8.45}$$

and hence

$$\exp(\alpha\widetilde{Y}) = \exp\left(\alpha\widetilde{Y}_0 + \int \frac{1}{2}\lambda^2\,d\,[M] + \alpha\int \widetilde{Z}\,\frac{dS}{S}\right)\mathcal{E}(\alpha\widetilde{L})$$

$$= \frac{\mathcal{E}(-\int \lambda\,dM + \alpha\widetilde{L})}{\exp(-\alpha\widetilde{Y}_0 - \int (\alpha\widetilde{Z} + \lambda)\,dS/S)}. \tag{8.46}$$

Comparing (8.46) with (8.44) motivates the following proposition which provides an explicit characterisation of the solution to the BSDE related to the utility maximisation problem without the claim.

Proposition 8.13. *Characterisation of Q^0 in terms of the BSDE (8.24).* *Let $(\widetilde{Y},\widetilde{Z},\widetilde{L})$ be the unique solution to the BSDE (8.24) in the sense outlined in Definition 8.7, with terminal condition $\xi = 0$ and generator $f(t,z) = -z\lambda_t\sigma_t^2 - \sigma_t^2\lambda_t^2/2\alpha$. Then,*

(i) The density process Z^0 of the entropy measure Q^0 is given by

$$Z^0 = \mathcal{E}\left(-\int \lambda\,dM + \alpha\widetilde{L}\right). \tag{8.47}$$

(ii) The density of the entropy measure Q^0 can be represented as

$$Z_T^0 = \exp\left(c_0 + \int_0^T \vartheta_t^0\,dS_t\right) \tag{8.48}$$

where $c_0 = -\alpha\widetilde{Y}_0$ and $\vartheta^0 = -\left(\alpha\widetilde{Z}_s + \lambda_s\right)/S$.

Proof.   It is sufficient to verify the conditions of Proposition 7.17 to hold for the candidate measure $\overline{Q}$ defined via the density (8.48). As the solution to the BSDE (8.24) is adapted, $\widetilde{Y}_0$ is a constant and the density of $\overline{Q}$ relative to P has the required form. Secondly, since $\widetilde{Y}_T = 0$ it follows from (8.46) that

$$Z_T^0 = \exp\left(c_0 + \int_0^T \vartheta_t^0\, dS_t + \alpha \widetilde{Y}_T\right) = \mathcal{E}\left(-\int \lambda\, dM + \alpha \widetilde{L}\right)_T.$$

Since $\widetilde{L} \in \mathcal{M}^2(P)$ and (8.37) is assumed to hold, the density Z_T^0 defines an equivalent measure $\overline{Q}$. Furthermore, $\int dS/S$ is a martingale under $\overline{Q}$ and hence

$$H\left(\overline{Q}, P\right) = E\left[\frac{d\overline{Q}}{dP} \log \frac{d\overline{Q}}{dP}\right] = -\alpha\widetilde{Y}_0 + E_{\overline{Q}}\left[\int_0^T \vartheta_t^0\, dS_t\right].$$

The r.h.s. is finite since (8.37) and $\widetilde{Z} \in L^2(M)$ imply that the stochastic integral is a martingale and $\widetilde{Y}$ is bounded by assumption.

Finally, we require that there exists $\varepsilon > 0$ such that

$$E\left[\exp\left(\varepsilon \int_0^T \left(\vartheta_t^0\right)^2 d[S]_t\right)\right] < \infty. \tag{8.49}$$

The assumption (8.37) implies that $\int \lambda\, dM$ is a BMO-martingale and furthermore it was shown in the proof of Theorem 8.9 that $\int \widetilde{Z}\, dM$ is a BMO-martingale. Hence, as

$$\int \vartheta^0\, dS = -\int \left(\alpha\widetilde{Z} + \lambda\right) \frac{dS}{S},$$

and $[\int dS/S] = [M]$ the claim (8.49) follows from the John-Nirenberg inequality. $\qquad\square$

Propositions 8.12 and 8.13 indicate that by considering the duality results presented in Chapter 7, the problem of identifying the utility indifference price and hedge can be reduced to solving a single optimisation problem under the entropy measure Q^0. The next result shows that the utility indifference price is the solution to a BSDE under the entropy measure Q^0.

Theorem 8.14. BSDE for the utility indifference price process under Q^0. *The price process $\pi_t = \pi_t(B; \alpha)$ is the first component of the unique solution of a quadratic BSDE with terminal condition $\xi = B$ and generator $f(t, z) \equiv 0$ under the entropy measure Q^0. That is, π can be represented as*

$$\pi_t = B - \int_t^T \widehat{Z}_u\, dS_u - \int_t^T d\widehat{L}_u + \frac{\alpha}{2} \int_t^T d[\widehat{L}]_u \tag{8.50}$$

where $\int \widehat{Z}\, dS$ and $\widehat{L}$ are strongly orthogonal $BMO\left(Q^0\right)$-martingales.

Proof. (sketch) Firstly, as $\pi_T(B; \alpha) = B$, by definition the terminal condition is satisfied. Applying the martingale optimality principle to the problem (8.42) gives that $J^B = \exp\left(\alpha\left(\pi - \int \vartheta^* \, dS\right)\right)$ is a Q^0-martingale for some $\vartheta^* \in \Theta$. By Jensen's inequality applied to the convex function $-\log(x)$ we obtain that $\pi(B; \alpha)$ is a Q^0-super-martingale with canonical decomposition

$$\pi(B; \alpha) = \pi_0(B; \alpha) - A + M \tag{8.51}$$

where A is a predictable increasing process and M is a Q^0-local martingale. The local martingale M has KW decomposition $M = \int \varphi^\pi \, dS + \widehat{L}$ where $\widehat{L}$ and S are strongly orthogonal local Q^0-martingales. In fact, it is verified in Mania & Schweizer (2005), Proposition 7, that $\widehat{L}$ and $\int \varphi^\pi \, dS$ are actually $BMO(Q^0)$-martingales. From (8.42) observe that

$$\exp(\alpha \pi_t(B; \alpha)) = \operatorname*{ess\,inf}_{\vartheta \in \Theta} E_{Q^0}\left[\exp\left(\alpha\left(B - \int_t^T \vartheta_u \, dS_u\right)\right)\bigg| \mathcal{F}_t\right]$$

so we may conclude from Theorem 8.5 that the process

$$J^{(\vartheta)} = \exp\left(-\alpha\left(\int \vartheta \, dS - \pi(B; \alpha)\right)\right)$$

$$= \exp\left(\alpha \pi_0(B; \alpha) - \alpha A + \alpha \widehat{L} + \alpha \int (\varphi^\pi - \vartheta) \, dS\right)$$

is a Q^0-sub-martingale for all $\vartheta \in \Theta$ and a Q^0-martingale for some $\vartheta \in \Theta$. Applying the Itô formula to $J^{(\vartheta)}$ and using the dynamics of $\pi(B; \alpha)$ given in (8.51) we obtain

$$J^{(\vartheta)} = J_0^{(\vartheta)} + \alpha \int J^{(\vartheta)}\left(\frac{\alpha}{2} d[\widehat{L}] + \frac{\alpha}{2}(\varphi^\pi - \vartheta)^2 \, d[S] - dA\right)$$

$$+ \alpha \int J^{(\vartheta)} \, d\widehat{L} + \alpha \int (\varphi^\pi - \vartheta)^2 \, dS.$$

By Theorem 8.5, the final two terms are Q^0-martingales. To ensure that $J^{(\vartheta)} \geq 0$ is a Q^0-sub-martingale the process

$$\frac{\alpha}{2}[\widehat{L}] - A + \frac{\alpha}{2}\int (\varphi^\pi - \vartheta)^2 \, d[S]$$

must be an increasing process for each $\vartheta \in \Theta$ and zero for some $\vartheta^* \in \Theta$. Hence,

$$A = \frac{\alpha}{2}[\widehat{L}] + \operatorname*{ess\,inf}_{\vartheta \in \Theta} \frac{\alpha}{2}\int (\varphi^\pi - \vartheta)^2 \, d[S]. \tag{8.52}$$

If we can prove $A = \frac{\alpha}{2}[\widehat{L}]$ then the first part of the theorem will follow from the canonical decomposition (8.51). Consider a sequence of stopping times

$$\tau_n = \inf\left\{ t \geq 0 \,\middle|\, \left|\int_0^t \varphi_u^\pi \, dS_u\right| \geq n \right\}$$

which are such that $\tau_n \to T$ almost surely and define a collection of strategies using $\vartheta^n = \varphi^\pi \mathbb{I}_{(0,\tau^n]}(t)$ such that $\vartheta^n \in \Theta$ for each $n \in \mathbb{N}$ where the set Θ is defined in (8.38). The strategies ϑ^n do not achieve the minimum in (8.52) so

$$\operatorname*{ess\,inf}_{\vartheta \in \Theta} \frac{\alpha}{2} \int_0^t (\varphi_u^\pi - \vartheta_u)^2 \, d[S]_u \leq \frac{\alpha}{2} \int_0^t (\varphi_u^\pi - \vartheta_u^n)^2 \, d[S]_u$$

$$= \frac{\alpha}{2} \int_{\tau^n}^{t \vee \tau^n} (\varphi_u^\pi)^2 \, d[S]_u$$

which tends to zero as $n \to \infty$. Thus, $\vartheta^* = \varphi^\pi = \widehat{Z}$ and hence $A = \frac{\alpha}{2}[\widehat{L}]$, and therefore the utility indifference price process has been demonstrated to be a solution of the BSDE (8.50).   □

The final proposition in this chapter verifies that when the structure condition (8.37) holds, the utility indifference price and hedge derived in Section 8.4 using the BSDEs (8.24) and (8.18) coincide with those derived in Theorem 8.14.

Proposition 8.15. *Characterisation of the utility indifference price and hedge.* *Let* (Y, Z, L), $(\widetilde{Y}, \widetilde{Z}, \widetilde{L})$ *denote the unique solutions to the BSDEs (8.18) and (8.24) respectively, in the sense outlined in Definition 8.7, with terminal conditions* $\xi = B$ *and* $\xi = 0$ *respectively and generator* $f(t, z) = -z\lambda_t \sigma_t^2 - \sigma_t^2 \lambda_t^2 / 2\alpha$. *Then,*

(i) The density process of the measure Q^α *with respect to* P^α *is given by*

$$Z^\alpha = \mathcal{E}\left(-\int \lambda \, dM + \alpha L\right).$$

Moreover,

$$Z_T^\alpha = \exp\left(-\alpha Y_0 - \int_0^T (\alpha Z_s + \lambda_s) \frac{dS_t}{S_t}\right). \tag{8.53}$$

(ii) The solution to the BSDE $(\pi, \widehat{Z}, \widehat{L})$ *introduced in Theorem 8.14 can be characterised in terms of the BSDEs (8.18) and (8.24) as*

$$(\pi, \widehat{Z}, \widehat{L}) = \left(Y - \widetilde{Y}, Z + \frac{\lambda}{\alpha}, L - \widetilde{L}\right).$$

Proof. (i) Repeating the argument which lead to (8.44),

$$u\left(t, -B; \alpha\right) = -\exp\left(-\alpha v\left(t, -B; \alpha\right)\right) = -\frac{Z_t^\alpha}{\overline{Z}_t^\alpha},$$

where

$$\overline{Z}^\alpha = \exp\left(c_\alpha + \int \vartheta^\alpha \, dS\right).$$

Repeating the steps which lead to (8.46) we obtain

$$\exp\left(\alpha Y\right) = \frac{\mathcal{E}\left(-\int \lambda \, dM + \alpha L\right)}{\exp\left(-\alpha Y_0 - \int \left(\alpha Z + \lambda\right) \frac{dS}{S}\right)}.$$

The verification procedure in Proposition 8.13 can be used to show that the density of Q^α relative to P^α is (8.53).

(ii) We obtain from (8.42) that

$$\exp\left(\alpha \pi_t\right) Z_t^0 = \operatorname*{ess\,inf}_{\vartheta \in \Theta} E\left[\exp\left(\alpha\left(B - \int_t^T \vartheta_u \, dS_u\right)\right) Z_T^0 \middle| \mathcal{F}_t\right],$$

hence

$$\exp\left(\alpha \pi_t\right) Z_t^0 = \exp(-\alpha \widetilde{Y}_0) u\left(t, -B; \alpha\right). \tag{8.54}$$

Inserting the expressions for Z^0 from Proposition 8.13 and π from Theorem 8.14 into (8.54) we obtain

$$\exp\left(\alpha \pi_0 + \alpha \int \varphi^\pi \, dS\right) \mathcal{E}\left(-\int \lambda \, dM + \alpha\left(\widetilde{L} + \widehat{L}\right)\right)$$

$$= \exp\left(\alpha\left(Y_0 - \widetilde{Y}_0\right) + \int \left(\alpha Z + \lambda\right) \frac{dS}{S}\right) \mathcal{E}\left(-\int \lambda \, dM + \alpha L\right)$$

and comparing terms (ii) follows. $\square$

8.7   Notes and further reading

The approach of this chapter originates with Hu *et al.* (2005) who used BSDE to solve utility maximisation problems in a Brownian market. Theorem 8.9 can be extended to cover closed but not necessarily convex constraint sets by using a measurable selection argument, see Hu *et al.* (2005).

The same set of admissible strategies has been used by Morlais (2009a) to extend these results to cover continuous semi-martingales as well as by Becherer (2006) and Morlais (2009b) to allow certain asset price models based upon a discontinuous filtration. Mania & Schweizer (2005) present an alternative approach which uses the duality results in Chapter 7 to deduce that the utility indifference price solves a BSDE. Describing the minimal entropy measure as the solution to a BSDE is examined by Rouge & El Karoui (2000) and Mania *et al.* (2003). The last section of this chapter is inspired by the approach taken by Mania & Schweizer (2005) and presents observations about the role of the components of the BSDEs similar to that given in Becherer (2006).

A key technicality with BSDEs is existence, uniqueness and comparison theorems. The existence results for quadratic BSDEs used by Hu *et al.* (2005) in a Brownian setting is due to Kobylanski (2000). Briand & Hu (2006) have extended this result to quadratic BSDEs with unbounded terminal values and a complimentary uniqueness result can be found in Briand & Hu (2008). In a general semi-martingale framework, existence and uniqueness of quadratic BSDEs satisfying Definition 8.6 have been studied by Morlais (2009a) and in greater generality in Hu & Schweizer (2009).

Chapter 9

Optimal Martingale Measures

This chapter is a survey of the 'zoology' of optimal martingale measures, their relation to different concepts of optimal hedging, and, finally, their computation in various models. Nota bene: the notion of admissible strategy is local to each section.

9.1 Esscher measure

The Esscher measure is a martingale measure obtained by an exponential tilting procedure. It is a popular choice since it exists under mild assumptions and is often easy to compute. There are two versions of the Esscher measure, depending on whether the price process is taken to be an ordinary or stochastic exponential. In the case when the price process is modelled as the stochastic exponential of a Lévy process, the Esscher measure coincides with the entropy measure. For more general processes, the linear Esscher measure minimises the entropy-Hellinger distance.

9.1.1 *Geometric case*

We consider a price process of the form $S = S_0 \exp(X)$ for some semi-martingale X with $X_0 = 0$, and recall that X is **exponentially special** if $\exp(X)$ is a special semi-martingale.

Definition 9.1. A strategy $\vartheta \in L(X)$ is **admissible** if the stochastic integral process $\int \vartheta \, dX$ is exponentially special.

Proposition 9.2. *Exponential compensator.* *In the case when ϑ is admissible, there exists a unique predictable process $K^X(\vartheta)$ such that*

$$Z^\vartheta := \exp\left(\int \vartheta \, dX - K^X(\vartheta)\right) \tag{9.1}$$

is a local martingale. $K^X(\vartheta)$ is called the **exponential compensator** of $\int \vartheta \, dX$.

Proof.   See Jacod & Shiryaev (2003), Proposition II.8.29. $\square$

If X is a Lévy process satisfying (EM) with $\beta = \vartheta \in \mathbb{R}$, the exponential compensator $K^X(\vartheta)$ coincides with the cumulant generating function $\kappa(\vartheta)$. This follows by considering the exponential martingale $\exp(\vartheta X - \kappa(\vartheta) t)$, and thus Z^ϑ coincides with the Esscher transform as introduced in Chapter 4. Returning to the general case:

Definition 9.3. For any admissible strategy ϑ such that Z^ϑ is a martingale, define a probability measure Q^ϑ via

$$\frac{dQ^\vartheta}{dP} = Z_T^\vartheta.$$

We refer to Q^ϑ as the **Esscher transform** with respect to ϑ.

We are interested in identifying the particular admissible strategy θ and corresponding Esscher transform Q^θ such that Q^θ is a martingale measure for $\exp(X)$.

Proposition 9.4. Geometric Esscher measure. *Let θ as well as $\theta + 1$ be admissible strategies such that Z^θ is a martingale. We have that $\exp(X)$ is a local Q^θ-martingale if and only if*

$$K^X(\theta + 1) = K^X(\theta). \tag{9.2}$$

*In this case we call Q^θ the **geometric Esscher measure**.*

Proof.   We have that $\exp(X)$ is a local Q^θ-martingale if and only if $Z^\theta \exp(X)$ is a local P-martingale. As

$$Z^\theta \exp(X) = \exp\left(\int (\theta + 1) \, dX - K^X(\theta)\right),$$

the statement follows since the exponential compensator of $\int (\theta + 1) \, dX$ is unique and equals $K^X(\theta + 1)$. $\square$

In our context, it is justified to talk about 'the' Esscher measure since by Theorem 4.2 of Kallsen & Shiryaev (2002), the Esscher measure for a univariate X is unique.

Proposition 9.5. Geometric Lévy processes. *Let $S = S_0 \exp(X)$ where X is a Lévy process, and let (γ, σ^2, ν) be the characteristic triplet* *of X relative to the truncation function h. Then there is at most one $\theta \in \mathbb{R}$ such that Q^θ defined by*

$$\frac{dQ^\theta}{dP} = \exp\left(\theta X_T - \kappa\left(\theta\right) T\right)$$

is the Esscher measure for S. In the case when it exists, θ satisfies

$$\gamma + \theta\sigma^2 + \frac{\sigma^2}{2} + \int \left(e^{\theta x}\left(e^x - 1\right) - h\left(x\right)\right) \nu\left(dx\right) = 0. \qquad (9.3)$$

Moreover, X is a Lévy process under Q^θ as well, with characteristic triplet

$$\gamma^\theta = \gamma + \theta\sigma^2 + \int \left(e^{\theta x} - 1\right) h\left(x\right) \nu\left(dx\right), \qquad (9.4)$$

$$\left(\sigma^2\right)^\theta = \sigma^2,$$

$$\nu^\theta\left(dx\right) = e^{\theta x} \nu\left(dx\right).$$

Proof. Note first that in the Lévy-case, the exponential compensator equals the cumulant generating function κ which has representation

$$\kappa\left(u\right) = \gamma u + \frac{\sigma^2}{2} u^2 + \int \left(e^{ux} - 1 - uh\left(x\right)\right) \nu\left(dx\right).$$

The equation $\kappa\left(\theta + 1\right) = \kappa\left(\theta\right)$ which determines the Esscher measure translates into (9.3). The last statement follows from Theorem 4.21 about structure preserving measure transforms, with Girsanov parameters $\beta = 1$, $y(x) = \exp\left(\theta x\right)$. $\qquad \square$

9.1.2 Linear case

Next we turn our attention to the case where the price process is modelled as $S = S_0 \, \mathcal{E}\left(X\right)$ for some semi-martingale X with $\Delta X > -1$ so that $\mathcal{E}\left(X\right)$ is strictly positive. As we have $dS = S_- \, dX$, S is a local martingale if and only if X is a local martingale. Similarly to the geometric case, we are looking for an admissible strategy θ such that X is a local martingale under the Esscher transform Q^θ.

A general discussion of the linear Esscher measure would require facts from semi-martingale characteristics which are, however, beyond the scope of this book; see Theorems 4.4, 4.5 of Kallsen & Shiryaev (2002) from which it also follows that the linear Esscher measure is unique. We first consider the case when X is a Lévy process.

Proposition 9.6. *Linear Esscher measure for Lévy processes.* Let X be a Lévy process with characteristic triplet (γ, σ^2, ν) relative to the truncation function h. Suppose there is $\theta \in \mathbb{R}$ with

$$E\left[|X_T| \exp(\theta X_T)\right] < \infty,$$

satisfying

$$\frac{d}{d\theta}\kappa(\theta) = 0 \tag{9.5}$$

which by the Lévy-Khintchine formula corresponds to the equation

$$\gamma + \theta\sigma^2 + \int \left(xe^{\theta x} - h(x)\right)\nu(dx) = 0. \tag{9.6}$$

Then X is a martingale under the measure Q^θ with density

$$\frac{dQ^\theta}{dP} = \exp\left(\theta X_T - \kappa(\theta)T\right).$$

Proof. Let $\theta \in \mathbb{R}$ be such that the Esscher transform Q^θ exists. X has then characteristic triplet with respect to Q^θ as in (9.4). It follows from the Lévy-Itô decomposition that X is a Q^θ-martingale if and only if $\gamma^\theta + (x - h(x)) * \nu^\theta = 0$ which, in terms of the characteristic triplet with respect to P, translates into equation (9.6). $\qquad \square$

Secondly, let X be a continuous semi-martingale, with canonical decomposition

$$X = M + \int \lambda\, d[M],$$

since $\langle M \rangle = [M]$, and

$$\int_0^T \lambda_t^2\, d[M]_t < \infty.$$

Proposition 9.7. *Linear Esscher measure for continuous processes.* Let X be a continuous semi-martingale such that the structure condition holds. The linear Esscher measure is given by

$$\frac{dQ^\theta}{dP} = \mathcal{E}\left(-\int \lambda\, dM\right)_T,$$

provided that $\mathcal{E}\left(-\int \lambda\, dM\right)$ is a martingale.

Proof. To determine the exponential compensator for an admissible strategy θ, observe that

$$Z^\theta = \exp\left(\int \theta\, dX - K^X(\theta)\right) = \exp\left(\int \theta\, dM + \int \theta\lambda\, d[M] - K^X(\theta)\right)$$

is a local martingale if and only if

$$K^X(\theta) = \int \theta\lambda\, d[M] + \frac{1}{2}\int \theta^2\, d[M]. \tag{9.7}$$

By Girsanov's theorem, Z^θ therefore is the density of a martingale measure if and only if it is a martingale and

$$\theta = -\lambda. \tag{9.8}$$

This ends the proof. $\qquad\qquad\qquad\qquad\qquad\qquad\qquad\qquad\qquad\qquad$ □

Note that we could have reached (9.8) in analogy to (9.5) also by formally differentiating the expression in (9.7) with respect to θ and equating to zero.

Thus the linear Esscher measure in the continuous setting is identical to the minimal martingale measure to be discussed in Section 9.5, but in general different from the entropy measure as we shall see in Section 9.2.4.

Moreover, the linear Esscher measure can be interpreted as an optimal martingale measure since it is intricately linked to the Hellinger measure.

Definition 9.8. (*i*) Let N be a local martingale such that $N_0 = 0$ and $\Delta N \geq -1$. If the adapted non-decreasing process

$$V(N) := \frac{1}{2}\langle N^c\rangle + \sum\left((1 + \Delta N)\log(1 + \Delta N) - \Delta N\right)$$

is locally integrable, the **entropy-Hellinger process** $H^E(N, P)$ is the compensator of V with respect to P. That is, $H^E(N, P)$ is the unique FV predictable process $\widetilde{V}$ such that $V - \widetilde{V}$ is a local martingale.

(*ii*) Let $Z = \mathcal{E}(N)$ be the density process of a martingale measure Q. Then we denote

$$h^E(Q, P) = H^E(N, P).$$

If there exists a solution to

$$\min_{Q\in\mathcal{M}} h^E(Q, P), \tag{9.9}$$

it is referred to as the **minimal entropy-Hellinger martingale measure** or rather, the Hellinger measure. Furthermore we denote the density process of the measure which achieves the minimum (9.9) by Z^H.

Under general conditions, the linear Esscher measure minimises the entropy-Hellinger distance, see Choulli & Stricker (2006), Theorem 3.3. However, the proof of this result is beyond the scope of this book.

The density process of the Hellinger measure Z^H is stable with respect to stopping: for every stopping time $\tau \in [0, T]$, Z_τ^H is the density of the Hellinger measure on $\mathcal{F}_\tau$. This is an immediate consequence of its interpretation as linear Esscher measure and the optional sampling theorem as we can write $Z^H = Z^\theta$, for some admissible strategy θ, such that the representation (9.1) as exponential compensator holds. In contrast, the density process Z^0 of the entropy measure need not be stable with respect to stopping. However, this is not surprising since Z^0 is linked via duality to the optimal strategy of a global utility maximisation problem on $[0, T]$.

9.2 Minimal entropy martingale measure

The entropy measure and its connection to asymptotic exponential utility indifference pricing and hedging have been discussed in Chapter 7. Our objective here is to give a general approach to calculating the density process of the entropy measure. The idea is to compare the canonical representation of the density of the entropy measure with the general form of the density process of a martingale measure at maturity. This will lead to the **optimal martingale measure equation** which can be solved in some special cases like exponential Lévy or orthogonal volatility models.

9.2.1 *Optimal martingale measure equation*

Let us assume that the filtration $\mathbb{F} = (\mathcal{F}_t)_{0 \leq t \leq T}$ fulfills the usual conditions and is generated by a one-dimensional Lévy process Y with Lévy measure ν where μ^Y and $\nu^Y(dx, dt) = \nu(dx)dt$ denote the jump measure and compensator, respectively. The continuous martingale part is a Brownian motion B modulo a multiplicative constant, which is assumed to be equal to one.

We denote by S an $\mathbb{F}$-adapted, locally bounded semi-martingale satisfying the structure condition with canonical decomposition

$$S = M + \int \lambda \, d\langle M \rangle,$$

$$K_T = \int_0^T \lambda_s^2 \, d\langle M \rangle_s < \infty.$$

(9.10)

By the weak representation property, we may write M as

$$M = \int \sigma^M dB + W^M(x) * (\mu^Y - \nu^Y) \qquad (9.11)$$

where $\sigma^M \in L^2_{loc}(B)$ and $W^M \in L^2_{loc}(\mu^Y)$.

Let us restate a criterion which must be fulfilled for a martingale measure to coincide with the entropy measure.

Proposition 9.9. Characterisation of entropy measure. *Assume there exists $Q \in \mathcal{M}^e$ with $H(Q, P) < \infty$. An equivalent martingale measure Q^* with finite relative entropy is the entropy measure if and only if a constant c and $\phi \in L(S)$ exist such that*

$$\frac{dQ^*}{dP} = \exp\left(c + \int_0^T \phi_t \, dS_t\right) \qquad (9.12)$$

with $E_Q[\int_0^T \phi_t \, dS_t] = 0$ for all $Q \in \mathcal{M}^e$ with finite relative entropy.

Based on this result, we will pursue the following strategy to determine the entropy measure: Firstly identify a candidate measure Q^* which can be represented as in (9.12). Then, to verify that Q^* is the entropy minimiser, show that:

(1) Q^* is an equivalent martingale measure;
(2) $H(Q^*, P) < \infty$;
(3) $\int \phi \, dS$ is a true Q-martingale for all $Q \in \mathcal{M}^e$ with finite relative entropy.

Although we will often skip this verification procedure and refer instead to the literature, each point can be approached as follows:

ad (1) Write the density process Z of Q^* as a stochastic exponential of the form

$$Z = \mathcal{E}\left(-\int \lambda \, dM + L\right) \qquad (9.13)$$

where L and $[M, L]$ are local P-martingales. It then suffices to show that this stochastic exponential is a true martingale, and then to apply Theorem 2.42(ii).

ad (2) This estimate involves calculating the Q^*-dynamics of the processes by invoking Girsanov's theorem.

ad (3) This may be shown by applying Lemma 7.16.

The main step consists of finding the candidate measure. Let us write the local martingale L by the weak representation property in the following way:

$$L = \int \sigma^L dB + W^L(x) * (\mu^Y - \nu^Y), \qquad (9.14)$$

for some $\sigma^L \in L^2_{loc}(B)$, $W^L \in L^2_{loc}(\mu^Y)$. Therefore, by (9.11),

$$[M, L] = \int \sigma^M \sigma^L \, ds + W^M(x)W^L(x) * \mu^Y.$$

Furthermore, the predictable bracket process

$$\langle M, L \rangle = \int \sigma^M \sigma^L \, ds + W^M(x)W^L(x) * \nu^Y$$

exists since M is locally bounded, and equals zero since $[M, L]$ is a local martingale. Therefore, as $\nu^Y(dx, dt) = \nu(dx)\, dt$ we get the orthogonality relation

$$\sigma^M \sigma^L + \int W^M(x)W^L(x)\, \nu(dx) = 0. \qquad (9.15)$$

Theorem 9.10. *Optimal martingale measure equation.* *Assume that there is a martingale measure Q^* such that (9.12) and (9.13) hold. Moreover, assume that $\left|\phi W^M(x)\right| * \nu^Y$ and $\left|\lambda W^M(x) - W^L(x)\right| * \nu^Y$ are of locally integrable variation. Then the quadruple $\left(c, \phi, \sigma^L, W^L\right)$ satisfies the equation*

$$c = -\int_0^T \left[\frac{1}{2}(\sigma_t^L - \lambda_t \sigma_t^M)^2 + \phi_t \lambda_t (\sigma_t^M)^2 \right.$$
$$\left. - \int \left(W_t^L(x) - (\phi_t + \lambda_t)W_t^M(x) + \phi_t \lambda_t (W_t^M(x))^2 \right) \nu(dx) \right] dt$$
$$+ \int_0^T \left(\sigma_t^L - (\phi_t + \lambda_t)\sigma_t^M \right) dB_t$$
$$+ \left((\log(1 - \lambda W^M(x) + W^L(x)) - \phi W^M(x)) * \mu^Y \right)_T. \qquad (9.16)$$

Proof. We apply Itô's formula to $\log Z$ to get, for $t \in [0, T]$, that

$$\log Z_t = \int_0^t \frac{1}{Z_{s-}}dZ_s - \frac{1}{2}\int_0^t \frac{1}{Z_{s-}^2}d\langle Z^c \rangle_s + \sum_{s \leq t} \left(\log Z_s - \log Z_{s-} - \frac{1}{Z_{s-}}\Delta Z_s \right).$$

By (9.13), and using the formula for the stochastic exponential, it follows that

$$\log Z_t = -\int_0^t \lambda_s \, dM_s + L_t - \frac{1}{2} \int_0^t \lambda_s^2 \, d\langle M^c \rangle_s + \int_0^t \lambda_s \, d\langle M^c, L^c \rangle_s - \frac{1}{2} \langle L^c \rangle_t$$
$$+ \sum_{s \le t} \left(\log \frac{Z_s}{Z_{s-}} + \Delta \int_0^s \lambda_u \, dM_u - \Delta L_s \right).$$

Plugging in the weak representations (9.11), (9.14), we reach

$$\log Z = \int (\sigma^L - \lambda \sigma^M) \, dB - \frac{1}{2} \int (\lambda \sigma^M - \sigma^L)^2 \, ds$$
$$+ (W^L(x) - \lambda W^M(x)) * (\mu^Y - \nu^Y)$$
$$+ (\log(1 - \lambda W^M(x) + W^L(x)) + \lambda W^M(x) - W^L(x)) * \mu^Y.$$

The integrability condition for $\lambda W^M(x) - W^L(x)$ implies by Lemma 4.14 that the r.h.s. equals

$$\int (\sigma^L - \lambda \sigma^M) \, dB - \frac{1}{2} \int (\lambda \sigma^M - \sigma^L)^2 \, ds$$
$$+ \left(\lambda W^M(x) - W^L(x) \right) * \nu^Y$$
$$+ \log(1 - \lambda W^M(x) + W^L(x)) * \mu^Y.$$

On the other hand, by (9.12) we have at the time horizon that

$$\log Z_T = c + \int_0^T \phi_t \, dS_t$$
$$= c + \int_0^T \phi_t \sigma_t^M \, dB_t + \left(\phi W^M(x) * (\mu^Y - \nu^Y) \right)_T$$
$$+ \int_0^T \left(\phi_t \lambda_t (\sigma_t^M)^2 + \phi_t \lambda_t \int (W_t^M(x))^2 \, \nu(dx) \right) dt.$$

Furthermore, the integrability condition for ϕW^M implies by Lemma 4.14 that we can write

$$(\phi W^M(x)) * (\mu^Y - \nu^Y) = (\phi W^M(x)) * \mu^Y - (\phi W^M(x)) * \nu^Y.$$

Equation (9.16) now follows by combining these equations. □

The optimal martingale measure equation (9.16) is an equation between random variables, and need not be satisfied prior to T. Furthermore, either σ^L or W^L can be eliminated by the constraint (9.15). Finally, the r.h.s. of (9.16) consists of three random variables of different types, namely a

dt-integral, a dB-integral, and a jump term, all of which have to add up to a constant P-a.s.

The optimal martingale measure equation gives only a necessary condition. Although it might be difficult to identify all solutions to (9.16), given the uniqueness of the entropy measure it suffices for our purposes to find just one solution (c, ϕ, σ^L, W^L) such that the verification procedure as outlined above can be carried out. The density process of the entropy measure is then given by (9.11), (9.13), and (9.14) as

$$Z = \mathcal{E}\left(-\int(\lambda\sigma^M - \sigma^L)dB - (\lambda W^M(x) - W^L(x)) * (\mu^Y - \nu^Y)\right). \quad (9.17)$$

9.2.2 Exponential Lévy case

Consider an exponential Lévy price process,

$$\frac{dS}{S_-} = \gamma\, dt + \sigma\, dB + d(x * (\mu^Y - \nu^Y)),$$

where we assume that $\sigma > 0$, and that $|x| * \nu^Y$ is of integrable variation. We have that λ in (9.10) is given by

$$\lambda S_- = \frac{\gamma}{\sigma^2 + \int x^2\, \nu(dx)}. \quad (9.18)$$

The functions σ^M and $W^M(x)$ in the optimal martingale measure equation correspond to σS_- and $x S_-$, respectively. Moreover, we set $\widehat{\lambda} := \lambda S_-$ and $\widehat{\phi} := \phi S_-$.

To find a solution to equation (9.16), we work with the *ansatz* that the jump term adds a constant contribution c to the r.h.s. That is, the integrand before the $*$-operator vanishes when W^L is taken as

$$W^L(x) = \widehat{\lambda}\, W^M(x) - 1 + e^{\widehat{\phi} W^M(x)}. \quad (9.19)$$

Similarly, the stochastic integral term is zero when

$$\sigma^L = (\widehat{\phi} + \widehat{\lambda})\sigma^M. \quad (9.20)$$

Thus, (9.20) together with (9.15) gives

$$\widehat{\phi} = -\widehat{\lambda} - \frac{\int W^M(x)W^L(x)\nu(dx)}{(\sigma^M)^2}. \quad (9.21)$$

From (9.18), (9.19) and (9.21) we finally get that, for all $t \geq 0$,

$$\gamma + \phi_t\sigma^2 + \int\left(xe^{\phi_t x} - x\right)\nu(dx) = 0, \quad (9.22)$$

given that the integral exists, which in particular shows that $\phi_t \equiv \phi$ must be a constant. When a solution to (9.22) exists, the verification procedure can be successfully carried out (see Rheinländer & Steiger (2006) for details), and we have found the entropy measure via its density

$$\frac{dQ^0}{dP} = \exp\left(c + \phi S_T\right).$$

Remark. The local boundedness of S is not needed for these results to hold. For this extension and a conceptional explanation of why the entropy measure preserves the Lévy property, see Esche & Schweizer (2005).

Note that by (9.6) and (9.22), for Lévy processes the entropy measure and the linear Esscher measure coincide. Hence the entropy measure is structure preserving, with Girsanov parameters $\beta = 1$ and $y(x) = \exp(\phi x)$. The density process of the entropy measure is, according to Theorem 4.21,

$$Z = \mathcal{E}\left(\int \phi \sigma \, dB + (e^{\phi x} - 1) * (\mu^Y - \nu^Y) \right).$$

9.2.3 *Orthogonal volatility case*

We assume that the filtration $\mathbb{F}$ is generated by a Lévy process Y. Let us consider on $[0, T]$ the stochastic volatility model

$$\frac{dS_t}{S_t} = \eta(t, V_t) \, dt + \sigma(t, V_t) \, dB_t \tag{9.23}$$

$$dV_t = \eta^V(t, V_t) \, dt + d\left(x * \mu^Y\right)_t$$

where B is a Brownian motion which is assumed to be the continuous martingale part of Y. In particular, $W^M = 0$. Since the filtration $\mathbb{F}$ is generated by Y, we can immediately see from the orthogonality relation (9.15) that $\sigma^L = 0$. We assume that

(1) The coefficients η, η^V and σ are Lipschitz-continuous in t and differentiable in y with bounded continuous derivatives. Here y corresponds to the 'V'-coordinate. Furthermore, σ is positive, bounded, and uniformly bounded away from zero on $[0, T] \times \mathbb{R}_+$.
(2) We have that $\widehat{\lambda} = \eta/\sigma^2$ is uniformly bounded on $[0, T] \times \mathbb{R}_+$.
(3) $\int |x| \, \nu(dx) < \infty$, where ν denotes the Lévy measure of Y.

These assumptions guarantee that there exists a unique solution to the SDE (9.23) which does not explode on $[0, T]$. Let us now simplify the optimal martingale measure equation (9.16). We work with the *ansatz* that there exists a sufficiently smooth function $u = u(t, y)$ such that

$$\log\left(1 + W_t^L\right) = u(t, V_t) - u(t, V_{t-}). \tag{9.24}$$

In addition, for all $y \in \mathbb{R}$ we set

$$u(T, y) = 0. \tag{9.25}$$

Using this *ansatz* we can write by Itô's formula

$$\left(\log(1 + W^L(x)) * \mu^Y\right)_T = \sum_{0 < t \leq T} \{u(t, V_t) - u(t, V_{t-})\}$$

$$= -u(0, V_0) - \int_0^T \left(\frac{\partial}{\partial t} u(t, V_t) + \eta_t^V \frac{\partial}{\partial y} u(t, V_t)\right) dt.$$

We may therefore rewrite equation (9.16) as

$$c + u(0, V_0) = -\int_0^T \left[\frac{1}{2}\widehat{\lambda}_t^2(\sigma_t^M)^2 + \phi_t\widehat{\lambda}_t(\sigma_t^M)^2 + \frac{\partial}{\partial t}u(t, V_t) + \eta_t^V \frac{\partial}{\partial y}u(t, V_t)\right.$$

$$\left. + \int W_t^L(x)\,\nu(dx)\right] dt$$

$$+ \int_0^T (\phi_t + \widehat{\lambda}_t)\sigma_t^M \, dB_t. \tag{9.26}$$

Setting the dB-term to zero, we get

$$\phi = -\widehat{\lambda}. \tag{9.27}$$

By (9.24), u determines W^L as

$$W_t^L = \exp\{u(t, V_t) - u(t, V_{t-})\} - 1. \tag{9.28}$$

Moreover, setting the dt-term to zero we reach

$$\frac{\partial}{\partial t}u(t, y) + \eta_t^V \frac{\partial}{\partial y}u(t, y) - \frac{1}{2}\widehat{\lambda}_t^2\left(\sigma_t^M\right)^2 + \int W_t^L(x)\,\nu(dx) = 0,$$

$$u(T, y) = 0.$$

Using the transformation $v(t, y) = \exp u(t, y)$ we derive the linear PDE

$$\frac{\partial}{\partial t}v(t, y) + \eta_t^V \frac{\partial}{\partial y}v(t, y) - \frac{1}{2}\widehat{\lambda}_t^2\left(\sigma_t^M\right)^2 v(t, y)$$

$$+ \int \left(v(t, y + x) - v(t, y)\right)\nu(dx) = 0, \tag{9.29}$$

$$v(T, y) = 1,$$

from which we can determine W^L, σ^L and hence the entropy measure via

$$W_t^L = \frac{v(t, V_t)}{v(t, V_{t-})} - 1, \tag{9.30}$$

$$\sigma^L(t, V) = 0.$$

Indeed, the verification procedure can be carried out as in Rheinländer & Steiger (2006). The density process of the entropy measure Q^0 is specified as in (9.17). Hence by Girsanov's theorem, the Lévy measure ν^0 under Q^0 is given as

$$\nu^0(dx) = \left(W^L(x) + 1 \right) \nu(dx).$$

As W^L need not be deterministic, the entropy measure is in general not structure preserving, that is Y need not be a Lévy process under Q^0.

9.2.4 *Continuous SV models*

In this subsection, we work with a filtration $\mathbb{G}$ such that all $\mathbb{G}$-martingales are continuous. In particular, for every $\mathbb{G}$-martingale M we have that $\langle M \rangle = [M]$, and for every continuous process A of finite variation that $[A] = 0$. Assuming the existence of a martingale measure for the positive price process S, the structure condition then states that there exists a local martingale M and a predictable process λ such that

$$S = M + \int \lambda \, d[M], \tag{9.31}$$

$$K_T = \int_0^T \lambda_t^2 \, d[M]_t < \infty.$$

By Theorem 7.14, the density of the entropy measure Q^0 can necessarily be written as

$$\frac{dQ^0}{dP} = \exp\left(c + \int_0^T \eta_t \, dS_t \right) \tag{9.32}$$

for some constant c and some $\mathbb{G}$-predictable process η. We will now look for a candidate martingale measure which can be represented in this manner.

On the other hand, by Theorem 2.42, the density of every martingale measure Q can be written as

$$\frac{dQ}{dP} = \mathcal{E}\left(-\int \lambda \, dM - L\right)_T \tag{9.33}$$

$$= \exp\left(-\int_0^T \lambda_t \, dM_t - \frac{1}{2}K_T - L_T - \frac{1}{2}[L]_T\right)$$

$$= \exp\left(-\int_0^T \lambda_t \, dS_t + \frac{1}{2}K_T - L_T - \frac{1}{2}[L]_T\right),$$

where L is some continuous local $(P, \mathbb{G})$-martingale strongly orthogonal to M, with $L_0 = 0$. To identify a candidate measure with representation (9.32), we look for such a local martingale L, a constant c and a predictable process ψ such that

$$\frac{1}{2}K_T = c + \int_0^T \psi_t \, dS_t + L_T + \frac{1}{2}[L]_T. \tag{9.34}$$

The optimal martingale measure equation (9.34) is an equation between random variables, not processes. This approach usually leads to a non-trivial martingale representation problem due to the presence of the quadratic variation $[L]$. Having found c, ψ, L such that (9.34) is satisfied, we have to check whether the random variable $\exp(c + \int_0^T \eta_t \, dS_t)$ with $\eta = \psi - \lambda$ is the density of a martingale measure with finite relative entropy and whether or not this martingale measure is really the entropy minimiser.

Returning to the optimal martingale measure equation (9.34), observe that it can be related to the solution of a quadratic BSDE denoted $(Y, Z, \widetilde{L})$, as defined in Definition 8.6 with $\alpha = 1$. We set the terminal condition $\xi = 0$ so that the first component of the solution, Y, can be written as

$$Y = Y_0 - \int f(t, Z) \, d[M] + \int Z \, dM + \widetilde{L} - \frac{1}{2}d[\widetilde{L}] \tag{9.35}$$

with $Y_T = 0$. Setting $L = -\widetilde{L}$ and inserting both (8.11) and the canonical decomposition (9.31) into the optimal martingale measure equation (9.34) gives

$$\frac{1}{2}K_T = c + Y_0 + \int_0^T (\psi_t + Z_t) \, dS_t - \int_0^T (f(t, Z_t) + \lambda_t Z_t) \, d[M]_t.$$

The $d[M]$-integrals cancel out when the generator is taken to be

$$f(t, z) = -\frac{1}{2}\lambda_t^2 - \lambda_t z. \tag{9.36}$$

Therefore, a candidate solution to the optimal martingale measure equation is $c = -Y_0$, $\psi = -Z$ and $L = -\tilde{L}$. Under the assumption that K_T is bounded, the entropy measure can indeed be written as

$$\frac{dQ^0}{dP} = \exp\left(-Y_0 - \int_0^T (Z_t + \lambda_t)\, dS_t\right),$$

and the verification procedure is presented in Proposition 8.13.

As a specific example consider the continuous orthogonal volatility model

$$\frac{dS_t}{S_t} = \eta(t, V_t)\, dt + \sigma(t, V_t)\, dB_t,$$

$$dV_t = \eta^V(t, V_t)\, dt + \sigma^V(t, V_t)\, dW_t$$

where η, η^V are bounded functions; σ, σ^V are bounded away from zero and B, W are orthogonal Brownian motions. To identify a candidate for the entropy measure, we apply the method described above with $dM = \sigma(t, V)\, S\, dB$ and $\lambda = \eta(t, V)/\sigma^2(t, V)\, S$. The dependence of the parameters on (t, v) will be suppressed for notational convenience.

Denote by $(Y, Z, \tilde{L})$ the solution to a quadratic BSDE with terminal condition $\xi = 0$ and generator f defined in (9.36). The orthogonal martingale L takes the form $L = -\int \varphi\, dW$ for some predictable integrand φ since the density process of every martingale measure Q for S can be written as

$$Z = \mathcal{E}\left(-\int \frac{\eta}{\sigma}\, dB - \int \varphi\, dW\right).$$

The first component of the solution to the BSDE, denoted Y, satisfies

$$Y = Y_0 + \int\left(\frac{1}{2}\frac{\eta^2}{\sigma^2} + \eta S Z\right) dt + \int \sigma S Z\, dB + \int \varphi\, dW - \frac{1}{2}\int \varphi^2\, dt \quad (9.37)$$

with $Y_T = 0$. In this setting, a candidate solution to the optimal martingale measure equation is $c = -Y_0$, $\psi = -Z/S$ and $L = -\int \varphi\, dW$. In fact, it is possible to derive a semi-explicit solution to the BSDE (9.37) by considering the exponential transform $U = \exp(Y)$. The process U has linear dynamics

$$dU = \frac{1}{2}\frac{\eta^2}{\sigma^2} U\, dt + ZU\, dS + \varphi U\, dW$$

with terminal condition $U_T = 1$. Applying Proposition 8.10 gives that

$$U_t = E_{\widehat{P}}\left[\exp\left(-\frac{1}{2}\int_t^T \frac{\eta^2}{\sigma^2}\, du\right)\middle| \mathcal{F}_t\right]$$

where $\widehat{P}$ denotes the martingale measure with density

$$\frac{d\widehat{P}}{dP} = \mathcal{E}\left(-\int \frac{\eta}{\sigma}\,dB\right)_T.$$

The process $U \exp\left(\int \eta^2/\sigma^2\,du\right)$ is a square-integrable $\widehat{P}$-martingale which can be represented under $\widehat{P}$ as

$$U \exp\left(-\frac{1}{2}\int \frac{\eta^2}{\sigma^2}\,du\right) = \widehat{c} + \int \vartheta^B\,dS + \int \vartheta^W\,dW,$$

where $\vartheta^B \in L^2(S)$, $\vartheta^W \in L^2(W)$. However, the process on the l.h.s. is adapted to the filtration generated by W so $\vartheta^B = 0$. Comparing this representation with the dynamics of U, we obtain that $\left(\log U, 0, \int \vartheta^W/U\,dW\right)$ is the unique solution to (9.37). Therefore, the density of the entropy measure may be written as

$$\frac{dQ^0}{dP} = \exp\left(-\log U_0 - \int_0^T \lambda_t\,dS_t\right)$$

which is distinct from the linear Esscher measure, see Proposition 9.7.

A similar approach can be used in the case of stochastic volatility driven by a correlated Brownian motion. This case falls inside the scope of the discussion about basis risk in Section 8.5.2. Alternatively, one can link the optimal martingale measure equation via a Feynman-Kac representation to a PDE, see Hobson (2004) for details.

9.3   Variance-optimal martingale measure

The variance-optimal measure is related to mean-variance hedging *under the statistical measure* (as opposed to risk-neutral measure). In our view, it may be more pragmatic to hedge using a mean-variance criterion with respect to a risk-neutral measure, for reasons to be discussed below. However, the study of the variance-optimal measure and the associated mean-variance hedging has given many impulses to the development of key ideas in hedging in incomplete markets as well as stochastic analysis; see the *notes and further reading* at the end of this chapter. We have included this section mainly for comparative reasons. For a more detailed account, the reader is referred to Schweizer (2001).

Let S be a locally square-integrable semi-martingale, that is $S_t^* := \sup_{0 \le u \le t} |S_u|$ is locally square-integrable, modelling the discounted price process of a risky asset.

Definition 9.11. We denote by $\mathcal{M}_2^e$ the space of all equivalent martingale measures for S whose density is square-integrable with respect to P. A strategy $\vartheta \in L(S)$ is called **admissible** if $\int \vartheta \, dS$ is a Q-martingale for all $Q \in \mathcal{M}_2^e$, with $\int_0^T \vartheta_t \, dS_t \in L^2(P)$. The space of admissible strategies is denoted by Θ.

Definition 9.12. If there exists an element of $\mathcal{M}_2^e$ which minimises $\sqrt{1 + \text{Var}\,[dQ/dP]}$ $(= \|dQ/dP\|_{L^2(P)})$ over all $Q \in \mathcal{M}_2^e$, then it is referred to as the **variance-optimal martingale measure**, and denoted by $\widetilde{P}$.

For many interesting models, we shall see that $\widetilde{P}$ does not exist which limits the practical value of this section. At first sight, the situation seems to be reasonably tractable for continuous price processes, as the following result by Delbaen & Schachermayer (1996) shows.

Theorem 9.13. *Existence and structure of $\widetilde{P}$ in the continuous case.* *If S is continuous and $\mathcal{M}_2^e \neq \emptyset$, then $\widetilde{P}$ exists and is unique. Moreover, the process $\widetilde{Z}$, $\widetilde{Z}_t := E_{\widetilde{P}}[\frac{d\widetilde{P}}{dP}|\mathcal{F}_t]$, can be written as*

$$\widetilde{Z} = 1 + \int \widetilde{\zeta} \, dS$$

for some $\widetilde{\zeta} \in \Theta$. In particular, $\widetilde{Z}$ is continuous.

Unfortunately, the assumption $\mathcal{M}_2^e \neq \emptyset$ is not always satisfied. In some well-known models like the correlated Stein & Stein or Heston model, $\widetilde{P}$ does not exist if the time horizon T is too large: Heath *et al.* (2001), Hobson (2004), whereas the minimal entropy martingale measure always exists in these models: Hobson (2004), Rheinländer (2005). Under the assumptions of the previous theorem, as $\widetilde{Z} > 0$ by equivalence, we can define $\widetilde{\beta} = \widetilde{\zeta}/\widetilde{Z}$, so that $\widetilde{Z}_T = \mathcal{E}(\int \widetilde{\beta} \, dS)_T$. Hence the density of the variance optimal measure may be expressed as stochastic exponential of a stochastic integral with respect to S. In contrast, the density of the entropy measure is given as ordinary exponential of a stochastic integral with respect to S.

Given a claim $H \in L^2(P)$, we now consider the following problem:

$$\min_{c,\vartheta} E\left[\left(H - c - \int_0^T \vartheta_t \, dS_t\right)^2\right],$$

where we minimise over all constants c and all $\vartheta \in \Theta$. When a solution exists, we call c and ϑ the mean-variance price and hedging strategy, respectively. In contrast to Section 6.1, this criterion is formulated under the real-world measure P.

If the price process is continuous and $\widetilde{P}$ exists, then it is possible to fully describe the mean-variance price and strategy, see Schweizer (2001).

Theorem 9.14. *Mean-variance hedging under P, continuous case.* *Suppose S is continuous and $\mathcal{M}_2^e \neq \emptyset$. Let $H \in L^2(P)$, and write the KW decomposition of its associated value process $\widetilde{V}$ under $\widetilde{P}$ as*

$$\widetilde{V} = \widetilde{V}_0 + \int \widetilde{\xi}\, dS + \widetilde{L}.$$

Then the mean-variance price is given as

$$\widetilde{V}_0 = E_{\widetilde{P}}\left[H\right], \tag{9.38}$$

and the mean-variance hedging strategy $\widetilde{\vartheta}$ is

$$\widetilde{\vartheta}_t = \widetilde{\xi}_t - \widetilde{\zeta}_t \left(\int_0^{t-} \frac{1}{\widetilde{Z}_u}\, d\widetilde{L}_u \right). \tag{9.39}$$

Notice that, in contrast to mean-variance hedging under the risk-neutral measure, the optimal strategy $\widetilde{\vartheta}$ does *not* equal the integrand $\widetilde{\xi}$ from the KW decomposition. On the other hand, the mean-variance price does have a similar structure, as it is the expectation under the variance-optimal martingale measure.

In the discontinuous case the situation is much less transparent. A counter-example in Schweizer (1996) shows that in general, the variance-optimal measure need not exist, or may only exist in a generalised form as a signed measure. An example of this pathological behaviour of the variance-optimal measure for exponential Lévy processes has been provided in Hubalek *et al.* (2006). In these examples there exist claims which have a positive payoff but a negative mean-variance price, hence this valuation method presents a financial paradox.

9.4   *q*-optimal martingale measure

For $q > 1$ denote by $\mathcal{M}_q^e$ the space of all equivalent martingale measures for S with densities with respect to P being in $L^q(P)$. The *q*-**optimal martingale measure** $Q^{(q)}$ is the unique solution of

$$\min_{Q \in \mathcal{M}_q^e} \left\| \frac{dQ}{dP} \right\|_{L^q(P)}. \tag{9.40}$$

The special case $q = 2$ is precisely the variance-optimal martingale measure. When S is continuous and $\mathcal{M}_q^e \neq \emptyset$, the *q*-optimal measure exists, see Grandits & Krawzcyk (1998). However, for $q \neq 2$ the q-optimal measure is not linked to any meaningful hedging concept. Interest into this measure comes from the fact that the entropy measure can be interpreted as the q-optimal measure for $q = 1$. This is exemplified by a result due to Grandits & Rheinländer (2002): if S is a continuous semi-martingale, and under some technical conditions, the measures $Q^{(q)}$ attaining (9.40) converge as $q \to 1$ in entropy to the minimal entropy measure Q^E, i.e.

$$H(Q^{(q)}, Q^E) \to 0.$$

An analogous result for exponential Lévy-processes has been provided in Jeanblanc *et al.* (2007). Hence the family of q-optimal measures interpolates between the entropy and the variance-optimal martingale measure. This property has been applied to the ordering of option prices, see Henderson (2004).

9.5 Minimal martingale measure

In this section, the price process S is assumed to be a locally square-integrable semi-martingale satisfying the structure condition. Recall from Theorem 2.42 that the general density process Z of a martingale measure for $S = M + \int \lambda \, d \langle M \rangle$ is a stochastic exponential of the form

$$Z = \mathcal{E}\left(- \int \lambda \, dM + L\right) \qquad (9.41)$$

where L and $[M, L]$ are local P-martingales. The converse statement, however, is not true. In particular, the choice $L = 0$ need not lead to the density of a martingale measure. This might be because the stochastic exponential $\widehat{Z} = \mathcal{E}(- \int \lambda \, dM)$ is not strictly positive, or that it fails to be a martingale on $[0, T]$. In the case when $\widehat{Z}$ is a strictly positive square-integrable martingale, the resulting measure is called the minimal martingale measure $\widehat{P}$. It is possible to write down the density of $\widehat{P}$ immediately after we have obtained λ and M in the canonical form (2.8). In particular, $\widehat{Z}$ can be determined directly from the price process data, so there is no need to use any optimisation procedure to find an appropriate process L in (9.41). This makes $\widehat{P}$ a very convenient choice, and consequently it enjoys some popularity. Furthermore, it has been linked to a locally risk-minimising hedging criterion, see Schweizer (1991). However, this approach is somewhat technical to state, so instead we present the concept of pseudo-optimal hedging, which in the one-dimensional case is essentially equivalent. This section is a brief survey of this circle of ideas, and the interested reader is referred to the detailed exposition Schweizer (2001) and the references contained therein.

Definition 9.15. Assume that $\widehat{Z} := \mathcal{E}(-\int \lambda\, dM) \in \mathcal{M}^2$, and is strictly positive. Then the measure $\widehat{P}$ with density

$$\frac{d\widehat{P}}{dP} = \mathcal{E}\left(-\int \lambda\, dM\right)_T$$

is called **minimal martingale measure**.

Note that $\widehat{P}$ is indeed a martingale measure which has already been observed in our discussion of Girsanov's theorem. Moreover, for continuous processes it coincides with the linear Esscher measure by Proposition 9.7.

Proposition 9.16. *Minimal measure and strongly orthogonal martingales.* *Assume that $\widehat{P}$ exists and let $L \in \mathcal{M}^2_{0,loc}(P)$ be strongly P-orthogonal to M, then $L \in \mathcal{M}^2_{0,loc}(\widehat{P})$.*

Proof. By integration by parts, and the stochastic differential equation satisfied by the stochastic exponential,

$$\widehat{Z}L = \int \widehat{Z}_-\, dL - \int L_-\widehat{Z}_-\lambda\, dM - \int \widehat{Z}_-\lambda\, d\,[M,L].$$

By assumption, all three integrators on the r.h.s. are local martingale integrators. This implies by Theorem 2.24 that the corresponding stochastic integral processes are local martingales as well. We conclude since L is a local $\widehat{P}$-martingale if and only if $\widehat{Z}L$ is a local P-martingale. □

Remark. Note that in general, if M and L are strongly P-orthogonal, S and L need not be strongly orthogonal under $\widehat{P}$.

In the following, strategies need not be self-financing. The cost process $C(\varphi)$ and mean self-financing strategies are defined as in Section 6.5 taking $A = 0$.

Definition 9.17. Let Θ_S denote the space of all strategies ϑ such that

$$E\left[\int_0^T \vartheta_t^2\, d\,[M]_t + \left(\int_0^T |\vartheta_t\, \lambda_t\, d\langle M\rangle_t|\right)^2\right] < \infty.$$

An L^2-strategy is a pair $\varphi = (\vartheta, \eta)$, where $\vartheta \in \Theta_S$ and η is adapted, such that the **value process**

$$V(\varphi) := \vartheta\, S + \eta$$

is cadlag, and $V_t(\varphi) \in L^2(P)$ for all $t \in [0, T]$. A strategy is **admissible** if it is mean self-financing and an L^2-strategy. Moreover, an admissible strategy is called **replicating** if $V_T(\varphi) = H$.

Definition 9.18. A replicating strategy φ is called **pseudo-optimal** if the martingale $C(\varphi)$ is strongly orthogonal to M.

The key idea in finding pseudo-optimal strategies is the following decomposition.

Definition 9.19. The random variable $H \in L^2(P)$ admits a **Föllmer-Schweizer (FS) decomposition** if it can be written as

$$H = H_0 + \int_0^T \xi_u^H \, dS_u + L_T^H \qquad (9.42)$$

where H_0 is a constant, $\xi^H \in \Theta_S$, $L^H \in \mathcal{M}_0^2(P)$ and L^H is strongly orthogonal to M.

The difference between the FS and KW decompositions is that in the FS decomposition, S is not necessarily a local P-martingale. An immediate consequence of the two preceding definitions is the following result which is proved in Schweizer (2001).

Theorem 9.20. *Existence of pseudo-optimal strategy.* Let $H \in L^2(P)$ admit an FS decomposition (9.42). Then there exists a pseudo-optimal strategy φ with $\vartheta = \xi^H$ and cost process $C(\varphi) = H_0 + L^H$. The associated value process equals $V(\varphi) = H_0 + \int \xi^H \, dS + L^H$.

Assume that the minimal measure $\widehat{P}$ exists. In the setting of the preceding theorem we have that $V_t(\vartheta) = E_{\widehat{P}}[H | \mathcal{F}_t]$ since the martingale property of L^H is preserved under $\widehat{P}$ by a non-local version of Proposition 9.16. In particular, the fair value associated with a pseudo-optimal strategy is the expectation of the claim under the minimal martingale measure.

We now assume that the FS decomposition exists and turn our attention to obtaining it. Actually, the situation is quite transparent in the continuous case.

Proposition 9.21. *Relationship between FS/KW decompositions, continuous case.* Suppose $H \in L^2(P)$ admits an FS decomposition; the minimal measure $\widehat{P}$ exists and that S is continuous. Then the FS decomposition is given by the KW decomposition (for $t = T$) under $\widehat{P}$ of the value process $V(\varphi)$ associated with the pseudo-optimal strategy φ.

Proof.   As $\widehat{P}$ is a martingale measure, $S \in \mathcal{M}^2_{loc}(\widehat{P})$ where local square-integrability follows from the continuity of S. Since $\langle S, L^H \rangle = \langle M, L^H \rangle = 0$, we have that, by uniqueness of the KW decomposition, (9.42) is the (local) KW decomposition of $V(\varphi)$ under $\widehat{P}$. $\square$

However, when S has jumps, in general the above correspondence no longer holds as is shown by the following result by Choulli *et al.* (2010) which is stated without proof.

Theorem 9.22. *Relationship between FS/KW decompositions, general case.* *Suppose $H \in L^2(P)$ admits an FS decomposition and that the minimal measure $\widehat{P}$ exists. Assume that the KW decomposition under $\widehat{P}$ of $V(\varphi) = H_0 + \int \xi^H \, dS + L^H$ exists and is given by*

$$V(\varphi) = V_0 + \int \xi \, dS + L,$$

with $L \in \mathcal{M}^2_{0,loc}(\widehat{P})$, strongly orthogonal to S. Then:

(i) *The process $\langle \widehat{Z}, [S, L] \rangle$ exists, and is absolutely continuous with respect to $\langle M \rangle$. We denote the corresponding Radon-Nikodym derivative by Φ.*

(ii) *$\Phi \in L(S)$, and $\xi^H = \xi - \Phi$, $L^H = L + \int \Phi \, dS$.*

Therefore, the FS and KW decompositions differ when Φ is non-zero which e.g. is the case if S is given as sum of a Brownian motion, a compensated Poisson process and a drift term, and H is a vanilla put option written on S (see Choulli *et al.* (2010)).

One problem with the minimal martingale measure is that it fails to exist in some situations where other popular optimal martingale measures exist. An example of this for continuous processes has been provided by Delbaen & Schachermayer (1998). Another example arises in a model for the price process of a defaultable asset. We assume that the price process $\widetilde{S}$ of a defaultable security modelled under P is of the form

$$d\widetilde{S}_t / \widetilde{S}_{t-} = a_t \, dt + b_t \, dB_t + c_t \, dN_t \tag{9.43}$$

where B is a Brownian motion and N is the counting process martingale associated to the one-jump process $H_t = \mathbb{I}_{\tau \leq t}$. We assume that H has a deterministic and continuous intensity μ which is bounded away from zero. For simplicity, the functions a, $b > 0$, and $c > -1 + \varepsilon$ for some $\varepsilon > 0$ are assumed to be deterministic and bounded but could be time-inhomogenous.

Our goal is now to calculate the minimal martingale measure for $\widetilde{S}$ or, equivalently, its stochastic logarithm $S = \int d\widetilde{S}/\widetilde{S}_-$. The semi-martingale decomposition of S may be written uniquely in the form

$$S = M + \int \lambda \, d\langle M \rangle \tag{9.44}$$

for a local martingale M and a predictable process λ. We assume that

$$K_T := \int_0^T \lambda_t^2 \, d\langle M \rangle_t = \int_0^T \frac{a^2}{b^2 + \mu c^2} \, dt < \infty.$$

It is readily computed from (9.43) that

$$dM = b \, dB + c \, dN, \quad \lambda = \frac{a}{b^2 + \mu c^2}.$$

Moreover, we have

$$\int \lambda \, dM = \int \frac{a}{b^2 + \mu c^2} \cdot (b \, dB + c \, dN),$$

$$\lambda \, \Delta M = \lambda c \, \Delta N,$$

so that the density of the minimal martingale measure is given by

$$\mathcal{E}\left(-\int \lambda \, dM\right)_T = \exp\left(-\int_0^T \lambda_t \, b_t \, dB_t - \frac{1}{2} \int_0^T \lambda_t^2 b_t^2 \, dt\right)$$

$$\times \prod_{0 < t \leq T} (1 - \lambda_t c_t \, \Delta N_t).$$

As there is one single jump of N with jump size one, the density of the minimal martingale measure becomes negative with non-zero probability when a jump occurs at $\tau \leq T$ and $\lambda_\tau c_\tau > 1$. Therefore, in general $\widehat{P}$ is merely a signed measure. This leads to a financial paradox: if one assigns to a claim H a price by taking the expectation under $\widehat{P}$, the claim represented by the indicator function of a set which is charged by $\widehat{P}$ with a negative value has a negative price even though it has a non-negative payoff. In this situation, we conclude that the minimal martingale measure is not a good choice. However, both linear Esscher and entropy measure exist and can be calculated, see Lee & Rheinländer (2011).

9.6   Notes and further reading

The Esscher measure has been discussed in full generality by Kallsen & Shiryaev (2002). It has been obtained in the Lévy case and for Barndorff-Nielsen & Shephard models by Hubalek & Sgarra (2006), (2009). The minimal entropy-Hellinger measure is studied in Choulli & Stricker (2005), (2006).

The entropy measure for exponential Lévy models has been computed in various degrees of generality by Chan (1999), Fujiwara & Miyahara (2003), Esche & Schweizer (2005) and Choulli & Stricker (2005), amongst others. The orthogonal volatility case was investigated by Benth & Meyer-Brandis (2005), whereas the approach via the optimal martingale measure equation is from Rheinländer & Steiger (2006) who also studied the entropy measure for general Barndorff-Nielsen & Shephard models with correlation. See also Rheinländer & Steiger (2010) for a related application to hedging with basis risk.

See Schweizer (1996) and Delbaen & Schachermayer (1996) for properties of the variance-optimal measure. Moreover, the closedness of spaces of stochastic integrals in $L^2(P)$ has been studied in Delbaen *et al.* (1997), and signed martingale measures and related $\mathcal{E}$-martingales are discussed in Choulli *et al.* (1998). The optimal mean-variance hedging strategy for continuous semi-martingales was obtained by Gourieroux, Laurent & Pham (1998) and Rheinländer & Schweizer (1997). More recently, Černý & Kallsen (2007) have extended these results to general semi-martingales, by using semi-martingale characteristics.

Key references for the minimal martingale measure and the Föllmer-Schweizer decomposition are Schweizer (1995) as well as Choulli *et al.* (2010). Various optimal martingale measures for defaultable asset price models are studied in Lee & Rheinländer (2010). Monoyios (2007) relates the entropy and minimal martingale measures via an Esscher transform in a two-factor incomplete market model.

Appendix A

Notation and Conventions

We usually work on a reference probability space $(\Omega, \mathcal{F}, P)$ and a finite interval $[0, T]$. Random variables and stochastic processes are always defined on Ω resp. $\Omega \times [0, T]$, unless otherwise stated. Processes are denoted by letters, hence X should be read as $X = (X_t)_{0 \leq t \leq T}$, whereas X_t is the random variable we get by evaluating X at time t. We denote by t also the deterministic process given by $t \mapsto t$, hence $X + t$ denotes a process and $X_t + t$ its evaluation at time t.

The expectation of a random variable B with respect to P is denoted by $E_P[B]$ or just $E[B]$ if there is no confusion about the choice of probability measure.

(In)equalities between random variables are always understood in the almost surely sense, and the reference to the measure (e.g. P-a.s.) is usually dropped unless this is important or not clear from the context. Equality between two processes is understood in the sense of indistinguishability: for two given stochastic processes X, Y the equality $X = Y$ is to be read as

$$P(X_t = Y_t \quad \text{for all } t \in [0, T]) = 1.$$

For a semi-martingale X, its left-continuous version is defined as $X_{t-} = \lim_{s \nearrow t} X_s$, and its jump process ΔX is given by $\Delta X_t = X_t - X_{t-}$. The term $\int_0^t \vartheta_s \, dX_s$ denotes the stochastic integral on the interval $(0, t]$; in particular, a jump of X at zero does not come into play.

Let τ be a stopping time, then X^τ denotes the stopped process given by $X_t^\tau = X_{t \wedge \tau}$.

> Standing assumptions appear in shadowboxes.

These assumptions hold throughout the section in which they are imposed.

Some frequently used symbols:

$\mathcal{M}$; $\mathcal{M}^e$; $\mathcal{M}^f$ denote the space of martingale measures which are absolutely continuous; equivalent; have finite relative entropy with respect to P.

$\mathcal{M}^2$; $\mathcal{M}^2_{loc}$; $\mathcal{M}^2_0$ denote the space of all square-integrable (s.i.) martingales; locally s.i. martingales; s.i. martingales M with $M_0 = 0$. For $M \in \mathcal{M}^2_{loc}$, $\mathcal{S}(M)$ denotes the stable subspace generated by M.

Let X be a semi-martingale. $L(X)$ denotes all X-integrable processes, so $\int \vartheta \, dX$ is well-defined for $\vartheta \in L(X)$.

Let M be a martingale. $L^2(M)$ is the space of all $\vartheta \in L(M)$ such that $\int \vartheta \, dM$ is a square-integrable martingale.

$\mathcal{E}(X)$ is the Doléans stochastic exponential of the semi-martingale X. Since $\mathcal{E}(X)$ depends on the whole path of X, we denote its value at time t by $\mathcal{E}(X)_t$ with no subscript 't' inside the bracket.

$[X^c]_t = [X]_t - \sum_{s \le t} (\Delta X_s)^2$, for a semi-martingale X.

$\mathbb{F}^X$ is the smallest augmented filtration $\mathbb{F}$ such that the stochastic process X is $\mathbb{F}$-adapted.

μ^X is the jump measure of an adapted cadlag process X, and ν^X denotes its predictable compensator. A $*$ indicates space-time integration, so for an integrable f

$$\left(f(x) * \mu^X \right)_t := \int_0^t \int f(x) \, \mu^X (ds, dx)$$

$\text{supp}(\mu)$ denotes the support of the measure μ.

For a vector $x \in \mathbb{R}^d$, $\|x\|$ denotes its Euclidean norm, and x' its transpose.

Bibliography

Ansel, J.P., Stricker, C. (1992) Lois de martingale, densités et décomposition de Föllmer Schweizer. *Annales de l'Institut Henri Poincaré* **28**, 375–392.

Ansel, J.P., Stricker, C. (1993) Decomposition de Kunita–Watanabe. *Séminaire de Probabilités* **XXVII**, 30–32.

Applebaum, D. (2009) *Lévy Processes and Stochastic Calculus*. 2nd edition. Cambridge University Press.

Azéma, J., Gundy, R.F., Yor, M. (1980) Sur l'integrabilité uniforme des martingales continues. *Séminaire de Probabilités* **XIV**, 53–61.

Barndorff-Nielsen, O.E., Shiryaev, A.N. (2011) *Change of Time and Change of Measure*. World Scientific.

Becherer, D. (2003) Rational hedging and valuation of integrated risks under constant absolute risk aversion. *Insurance, Mathematics and Economics* **33**, 1–28.

Becherer, D. (2006) Bounded solutions to backward SDE's with jumps for utility optimisation and indifference hedging. *Annals of Applied Probability* **16**, 2027–2054.

Beghdadi-Sakrani, S. (2003) Some remarkable pure martingales. *Annales Institute H. Poincaré* **39**, 287–299.

Benth, F.E., Benth, J.S., Koekebakker, S. (2008) *Stochastic Modelling of Electricity and Related Markets*. World Scientific.

Benth, F.E., Groth, M., Wallin, O. (2010) Derivative-free Greeks for the Barndorff-Nielsen and Shephard stochastic volatility model. *Stochastics* **82**, 291–313.

Benth, F.E., Meyer-Brandis, T. (2005) The density process of the minimal entropy martingale measure in a stochastic volatility model with jumps. *Finance and Stochastics* **9**, 563–575.

Biagini, F., Cretarola, A. (2009) Local risk minimisation for defaultable markets. *Mathematical Finance* **19**, 669–689.

Biagini, F., Rheinländer, T., Widenmann, J. (2011) Hedging mortality claims with longevity bonds. *Preprint*.

Biagini, S., Frittelli, M. (2008) A unified framework for utility maximisation problems: An Orlicz space approach. *Annals of Applied Probability* **18**, 929–966.

Biagini, S., Frittelli, M., Grasselli, M. (2011) Indifference price with general semi-martingales. *To appear in Mathematical Finance.*

Bielecki, T.R., Rutkowski, M. (2002) *Credit Risk: Modelling, Valuation and Hedging.* Springer.

Boyarchenko, S.I., Levendorskiĭ, S.Z. (2002) *Non-Gaussian Merton-Black-Scholes Theory.* World Scientific.

Briand, P., Hu, Y. (2006) BSDE with quadratic growth and unbounded terminal value. *Probability Theory and Related Fields* **136**, 604–618.

Briand P., Hu, Y. (2008) Quadratic BSDEs with convex generators and unbounded terminal conditions, *Probability Theory and Related Fields* **141**, 543–567.

Carmona, R. (Ed.) (2008) *Indifference Pricing: Theory and Applications.* Princeton University Press.

Carr, P., Geman, H., Madan, D.B., Yor, M. (2002). The fine structure of asset returns: An empirical investigation. *Journal of Business* **75**, 305–332.

Carr, P., Geman, H., Madan, D.B., Yor, M. (2003) Stochastic volatility for Lévy processes. *Mathematical Finance* **13**, 345–382.

Carr, P., Lee, T. (2009) Put-call symmetry: extensions and applications. *Mathematical Finance* **19**, 523–560.

Cerny, A.S., Shiryaev, A.N. (2002) Vector stochastic integrals and the fundamental theorems of asset pricing. *Proceedings of the Steklov Institute of Mathematics* **237**, 6–49.

Černý, A., Kallsen, J. (2007) On the structure of general mean-variance hedging strategies. *Annals of Probability* **35**, 1479–1531.

Chan, T. (1999) Pricing contingent claims on stocks driven by Lévy processes. *Annals of Applied Probability* **9**, 504–528.

Chou, C.S., Meyer, P.A., Stricker, C. (1980) Sur les intégrales stochastiques de processus prévisibles non bornés. *Séminaire de Probabilités* **XIV**, 128–139.

Choulli, T., Krawczyk, L., Stricker, C. (1998) $\mathcal{E}$-martingales and their applications in mathematical finance. *Annals of Probability* **26**, 853–876.

Choulli, T., Stricker, C. (2005) Minimal entropy-Hellinger martingale measure in incomplete markets. *Mathematical Finance* **15**, 465–490.

Choulli, T., Stricker, C. (2006) More on minimal entropy-Hellinger martingale measure. *Mathematical Finance* **16**, 1–19.

Choulli, T., Vandaele, N., Vanmaele, M. (2010) The Föllmer-Schweizer decomposition: comparison and description. *Stochastic Processes and their Applications* **120**, 853–872.

Cont, R. (2006) Model uncertainty and its impact on derivative instruments. *Mathematical Finance* **16**, 519–542.

Cont, R. (Ed.) (2010) *Encyclopedia of Quantitative Finance.* Wiley.

Cont, R., Tankov, P. (2003) *Financial Modelling with Jump Processes.* Chapman & Hall/CRC Press.

Cont, R., Tankov, P. (2004) Non-parametric calibration of jump-diffusion option pricing models. *Journal of Computational Finance* **7**, 1–49.

Cont, R., Tankov, P. (2006) Retrieving Lévy processes from option prices: Regularisation of an ill-posed inverse problem. *SIAM Journal on Control and Optimisation* **45**, 1–25

Cont, R., Tankov, P., Voltchkova, E. (2007) Hedging with options in presence of jumps. In: *Stochastic Analysis and Applications: The Abel Symposium 2005 in honor of Kiyosi Ito,* Benth, F.E., Di Nunno, G., Lindstrom, T., Øksendal, B., Zhang, T. (Eds.), Springer, 197–218.

Cont, R., Voltchkova, E. (2005) Integro-differential equations for option prices in exponential Lévy models. *Finance & Stochastics* **9**, 299–325.

Conway, J. B. (1990) *A Course in Functional Analysis.* 2nd edition. Springer.

Cox, A.M.G., Hobson, D.G. (2005) Local martingales, bubbles and option prices. *Finance and Stochastics* **9**, 477–492.

Csiszár, I. (1975) I-divergence geometry of probability distributions and minimisation problems. *Annals of Probability* **3**, 146–158.

Dahl, M., Melchior, M., Møller, T. (2008) On systematic mortality risk and risk-minimisation with survivor swaps. *Scandinavian Actuarial Journal* **2**, 114–146.

Davis, M. (1997) Option pricing in incomplete markets. In: M. Dempster and S. Pliska (Eds.), *Mathematics of Derivative Securities,* 216–226. Cambridge University Press.

Delbaen, F., Grandits, P., Rheinländer, T., Samperi, D., Schweizer, M., Stricker, C. (2002) Exponential hedging and entropic penalties. *Mathematical Finance* **12**, 99–123.

Delbaen, F., Monat, P., Schachermayer, W., Schweizer, M., Stricker, C. (1997) Weighted norm inequalities and hedging in incomplete markets. *Finance and Stochastics* **1**, 181–227.

Delbaen F., Schachermayer, W. (1995a) The existence of absolutely continuous martingale measures. *Annals of Applied Probability* **5**, 926–945.

Delbaen F., Schachermayer, W. (1995b) Arbitrage possibilities in Bessel processes and their relations to local martingales. *Probability Theory and Related Fields* **102**, 357–366.

Delbaen, F., Schachermayer, W. (1996) The variance-optimal martingale measure for continuous processes. Bernoulli **2**, 81–105.

Delbaen, F., Schachermayer, W. (1998) A simple counter-example to several problems in the theory of asset pricing, which arises in many incomplete markets. *Mathematical Finance* **8**, 1–12.

Delbaen F., Schachermayer, W. (2006) *The Mathematics of Arbitrage.* Springer.

Dellacherie, C., Meyer, P.A. (1980) *Probabilités et potentiel, Ch. V à VIII.* Hermann.

Dritschel, M., Protter, P. (1999) Complete markets with discontinuous security price. *Finance and Stochastics* **3**, 203–214.

Duffie, D., Filipovic, D., Schachermayer, W. (2003) Affine processes and applications in finance. *Annals of Applied Probability* **13**, 984–1053.

Eberlein, E., Keller, U. (1995) Hyperbolic distributions in finance. *Bernoulli* **1**, 281–299.

Eberlein, E., Raible, S. (2001) Some analytic facts on the generalised hyperbolic model. *In: Proceedings of the 3rd European Meeting of Mathematics,* Progress in Mathematics 202, C. Casacuberta, *et al.* (Eds.), Birkhäuser, 367–378.

Eberlein, E., Papapantoleon, A. (2005) Symmetries and pricing of exotic options in Lévy models. *In: Exotic option pricing and advanced Lévy models,* A. Kyprianou, W. Schoutens, P. Wilmott (Eds.), Wiley, 99–128.

Eberlein, E., Papapantoleon, A., Shiryaev, A.N. (2008). On the duality principle in option pricing: semimartingale setting. *Finance and Stochastics* **12**, 265–292.

El Karoui, N., Quenez, M.C. (1995) Dynamic programming and pricing of contingent claims in an incomplete market. *SIAM Journal on control and optimisation* **33**, 29–66.

Emery, M., Perkins, E. (1982) La filtration de $B + L$. *Zeitschrift für Wahrscheinlichkeitstheorie und verwandte Gebiete* **59**, 383–390.

Emery, M., Stricker, C., Yan, J.A. (1983) Valeurs prises par les martingales locales continues a un instant donne. *Annals of Probability* **11**, 635–641.

Esche, F., Schweizer, M. (2005) Minimal entropy preserves the Lévy property: How and why. *Stochastic Processes and their Applications* **115**, 299–327.

Fernholz, E.R., Karatzas I., Kardaras, C. (2005) Diversity and relative arbitrage in equity markets. *Finance and Stochastics* **9**, 1–27.

Föllmer, H., Kramkov, D. (1997) Optional decompositions under constraints. *Probability Theory and Related Fields* **109**, 1–25.

Föllmer, H., Leukert, P. (1999). Quantile hedging. *Finance and Stochastics* **3**, 251–273.

Föllmer, H., Leukert, P. (2000). Efficient hedging: Cost versus shortfall risk. *Finance and Stochastics* **4**, 117–146.

Föllmer, H., Sondermann, D. (1986) Hedging of non-redundant contingent claims. *In: Contributions to Mathematical Economics.* In Honor of G. Debreu (Eds. W. Hildenbrand and A. Mas-Colell), Elsevier Science Publications, 205–223.

Friedman, A. (1975) *Stochastic Differential Equations and Applications.* Vol. 1, Academic Press.

Frittelli, M. (2000) The minimal entropy martingale measure and the valuation problem in incomplete markets. *Mathematical Finance* **10**, 39–52.

Frittelli, M., Biagini, S., Scandolo, G. (2011) *Duality in Mathematical Finance.* Springer (forthcoming).

Fujiwara, T., Miyahara, Y. (2003) The minimal entropy martingale measures for geometric Lévy processes. *Finance and Stochastics* **7**, 509–531.

Geiss, C., Geiss, S. (2006) On an approximation problem for stochastic integrals where random time nets do not help. *Stochastic Processes and Applications* **116**, 407–422.

Geiss, S., Hujo, M. (2007) Interpolation and approximation in $L^2(\gamma)$. *Journal of Approximation Theory* **144**, 213–232.

Gerber, H.U. (1979) *An Introduction to Mathematical Risk Theory.* S.S. Huebner Foundation monograph.

Grandits, P., Krawczyk, L. (1998) Closedness of some spaces of stochastic integrals. *Séminaire de Probabilités* **XXXII**, 73–85.

Grandits, P., Rheinländer, T. (2002) On the minimal entropy martingale measure. *Annals of Probability* **30**, 1003–1038.

Grasselli, M.R., Hurd, T.R. (2007) Indifference pricing and hedging for volatility derivatives. *Applied Mathematical Finance* **14**, 303–317.

Goll, T., Rüschendorf, L. (2001) Minimax and minimal distance martingale measures and their relationship to portfolio optimisation. *Finance and Stochastics* **5**, 557–581.

Gossen-Dombrowsky, R. (1992) *Ein Ansatz zur Optionsbewertung in stochastischen Modellen ohne ein äquivalentes Martingalmaß*. Master's thesis, University of Bonn.

Gourieroux, C., Laurent, J.P., Pham, H. (1998) Mean-variance hedging and numeraire. *Mathematical Finance* **8**, 179–200.

He, S.W., Wang, J.G., Yan, J.A. (1992) *Semimartingale Theory and Stochastic Calculus*. CRC Press.

Heath, D., Platen, E., Schweizer, M. (2001) A comparison of two quadratic approaches to hedging in incomplete markets. *Mathematical Finance* **11**, 385–413.

Henderson, V. (2004) Analytical comparisons of option prices in stochastic volatility models. *Mathematical Finance* **15**, 49–59.

Hobson, D.G. (1998) Volatility mis-specification, option pricing and super-replication via coupling. *Annals of Applied Probability* **8**, 193–205.

Hobson, D.G. (2004) Stochastic volatility models, correlation, and the q-optimal measure. *Mathematical Finance* **14**, 537–556.

Hobson, D.G. (2010) Comparison results for stochastic volatility models via coupling. *Finance and Stochastics* **14**, 129–152.

Hobson, D.G., Rogers, L.C.G. (1998) Complete models with stochastic volatility. *Mathematical Finance* **8**, 27–48.

Hu, Y., Imkeller, P., Müller, M. (2005) Utility maximisation in incomplete markets. *Annals of Applied Probabability* **15**, 1691–1712.

Hu, Y., Schweizer, M. (2009) Some new BSDE results for an infinite-horizon stochastic control problem. Preprint, ETH Zurich.

Hubalek, F., Krawczyk, L., Kallsen, J. (2006). Variance–optimal hedging for processes with stationary independent increments. *Annals of Applied Probability* **16**, 853–885.

Hubalek, F., Sgarra, C. (2006) Esscher transforms and the minimal entropy martingale measure for exponential Lévy models. *Quantitative Finance* **6**, 125–145.

Hubalek, F., Sgarra, C. (2009) On the Esscher transforms and other equivalent martingale measures for Barndorff-Nielsen and Shephard stochastic volatility models with jumps. *Stochastic Processes and their Applications* **119**, 2137–2157.

Ihara, S. (1992) *Information Theory for Continuous Systems*. World Scientific.

Ilhan, A., Sircar, R. (2006) Optimal static-dynamic hedges for barrier options. *Mathematical Finance* **16**, 359–385.

Jacod, J. (1979) *Calcul Stochastique et Problèmes de Martingales*. LNM **714**, Springer.

Jacod, J., Shiryaev, A.N. (2003) *Limit Theorems for Stochastic Processes.* 2nd edition. Springer.

Jarrow, R., Protter, P. (2005) *Large Traders, Hidden Arbitrage, and Complete Markets.* Journal of Banking and Finance **29**, 2803–2820.

Jarrow, R., Protter, P., Shimbo, K. (2010) Asset price bubbles in incomplete markets. *Mathematical Finance* **20**, 145–185.

Jeanblanc, M., Kloeppel, S., Miyahara, Y. (2007) Minimal martingale measures for exponential Lévy processes. *Annals of Applied Probability* **17**, 1615–1638.

Jeanblanc, M., Yor, M., Chesney, M. (2009) *Mathematical Methods for Financial Markets.* Springer.

Jeulin, T., Yor, M. (1979) Inegalité de Hardy, semimartingales et faux-amis. *Séminaire de Probabilités* **XIII**, 332–359.

Kabanov, Y., Stricker, C. (2001) On equivalent martingale measures with bounded densities. *Séminaire de Probabilités* **XXXV**, 139–148.

Kabanov, Y., Stricker, C. (2002) On the optimal portfolio for the exponential utility maximisation: Remarks to the six-author paper. *Mathematical Finance* **12**, 125–134.

Kallsen, J. (2000) Optimal portfolios for exponential Lévy processes. *Mathematical Methods in Operations Research* **51**, 357–374.

Kallsen, J. (2006) A didactic note on affine stochastic volatility models. *In: From Stochastic Calculus to Mathematical Finance: The Shiryaev Festschrift,* Kabanov, Y., Lipster, R., Stoyanov, J. (Eds.), Springer, 343–368.

Kallsen, J., Pauwels, A. (2010a) Variance-optimal hedging in general affine stochastic volatility models. *Advances in Applied Probability* **42**, 83–105.

Kallsen, J., Pauwels, A. (2010b) Variance-optimal hedging for time-changed Lévy processes. *Applied Mathematical Finance,* 1–28.

Kallsen, J., Rheinländer, T. (2011) Asymptotic utility-based pricing and hedging for exponential utility. *To appear in Statistics and Decisions.*

Kallsen, J., Shiryaev, A.N. (2002) The cumulant process and Esscher's change of measure. *Finance and Stochastics* **6**, 397–428.

Kazamaki, N. (1994) *Continuous exponential martingales and BMO.* LNM 1579, Springer.

Kluge, W. (2005) *Time-inhomogeneous Lévy processes in Interest Rate and Credit Risk Models.* Dissertation, University of Freiburg.

Kobylanski, M. (2000) Backward stochastic differential equations and partial differential equations with quadratic growth. *Annals of Probability* **28**, 558–602.

Kramkov, D., Sîrbu, M. (2006) The sensitivity analysis of utility based prices and the risk-tolerance wealth processes. *Annals of Applied Probability* **16**, 2140–2194.

Kramkov, D., Sîrbu, M. (2007) Asymptotic analysis of utility based hedging strategies for a small number of contingent claims. *Stochastic Processes and Their Applications* **117**, 1606–1620.

Kyprianou, A. (2006) *Introductory Lectures on Fluctuations of Lévy Processes with Applications.* Springer.

Lee, Y., Rheinländer, T. (2011) Optimal martingale measures for defaultable assets. *Preprint, London School of Economics.*

Liptser, R. S., Shiryaev, A.N. (2000) *Statistics of Random Processes I.* 2nd edition, Springer.

Mania, M., Santacroce, M., Tevzadze R. (2003) A semimartingale BSDE related to the minimal entropy martingale measure. *Finance & Stochastics* **7**, 385–402.

Mania, M., Schweizer, M. (2005) Dynamic exponential utility indifference valuation. *Annals of Applied Probability* **15**, 2113–2143.

Miyahara, Y. (1996) Canonical martingale measures of incomplete assets markets. *In: Probability Theory and Mathematical Statistics: Proceedings of the Seventh Japan-Russia Symposium, Tokyo* (eds. S. Watanabe *et al.*), 343–352.

Molchanov, I., Schmutz, M. (2010) Multivariate extension of put-call symmetry. *SIAM Journal of Financial Mathematics* **1**, 398–426.

Møller, T. (1998) Risk-minimising hedging strategies for unit-linked life insurance contracts. *ASTIN Bulletin* **28**, 17–47.

Møller, T. (2001) Risk-minimising hedging strategies for insurance payment processes. *Finance and Stochastics* **5**, 419–446.

Monoyios, M. (2007) The minimal entropy measure and an Esscher transform in an incomplete market model. *Statistics and Probability Letters* **77**, 1070–1076.

Monoyios, M. (2010) Utility-based valuation and hedging of basis risk with partial information. *Applied Mathematical Finance* **17**, 519–551.

Morlais, M.A. (2009) Utility maximisation in a jump market model. *Stochastics* **81**, 1–27.

Morlais, M.A. (2009) Quadratic BSDEs driven by a continuous martingale and applications to the utility maximisation problem. *Finance and Stochastics* **13**, 121–150.

Neveu, J. (1975) *Martingales à temps discret.* Masson.

Ocone, D.L. (1993) A symmetry characterisation of conditionally independent increment martingales. *In: D. Nualart, M. Sanz (Eds.), Proceedings of the San Felice Workshop on Stochastic Analysis,* Birkhäuser.

Osterrieder, J., Rheinländer, T. (2006) Arbitrage opportunities in diverse markets via a non-equivalent measure change. *Annals of Finance* **2**, 287–301.

Pham, H. (2009) *Continuous-time stochastic control and optimisation with financial applications.* Springer.

Pham, H., Rheinländer, T., Schweizer, M. (1998) Mean-variance hedging for continuous processes: new proofs and examples. *Finance and Stochastics* **2**, 173–198.

Protter, P. (2001) A partial introduction to financial asset theory. *Stochastic Processes and their Applications* **91**, 169–203.

Protter, P. (2005) *Stochastic Integration and Differential Equations — A New Approach.* 2nd edition, version 2.1. Springer.

Protter, P., Shimbo, K. (2008) No arbitrage and general semimartingales. *In: Markov Processes and Related Topics: A Festschrift for Thomas G. Kurtz*

(Beachwood, Ohio, USA: Institute of Mathematical Statistics), 267–283.

Raible, S. (2000) *Lévy Processes in Finance: Theory, Numerics, and Empirical Facts.* Dissertation, University of Freiburg.

Revuz, D., Yor, M. (1998) *Continuous Martingales and Brownian Motion.* 3rd edition. Springer.

Rheinländer, T. (2005) An entropy approach to the Stein and Stein model with correlation. *Finance and Stochastics* **9**, 399–412.

Rheinländer, T., Schmutz, M. (2011) Quasi self-dual processes. *Preprint.*

Rheinländer, T., Schweizer, M. (1997) On L^2-projections on a space of stochastic integrals. *Annals of Probability* **25**, 1810–1831.

Rheinländer, T., Steiger, G. (2006) The minimal entropy martingale measure for general Barndorff-Nielsen/Shephard models. *Annals of Applied Probability* **16**, 1319–1351.

Rheinländer, T., Steiger, G. (2010) Utility indifference hedging with exponential additive processes. *Asia-Pacific Financial Markets* **17**, 151–169.

Roger, L.C.G., Williams, D. (2000) *Diffusions, Markov Processes and Martingales: Volume 2, Itô Calculus.* 2nd edition. Cambridge University Press.

Romano, M., Touzi, N. (1997) Contingent claims and market completeness in a stochastic volatility model. *Mathematical Finance* **7**, 399–412.

Rouge, R., El Karoui, N. (2000) Pricing via utility maximisation and entropy. *Mathematical Finance* **10**, 259–276.

Sato, K. (1999) *Lévy Processes and Infinitely Divisible Distributions.* Cambridge University Press.

Schachermayer, W. (2003) A super-martingale property of the optimal portfolio process. *Finance and Stochastics* **7**, 433–456.

Schmutz, M., Zürcher, T. (2010) Static replications with traffic light options. *Technical report, University of Bern.*

Schmutz, M., Zürcher, T. (2011) A Stieltjes approach to static hedging. *Technical report, University of Bern.*

Schoutens, W. (2003) *Lévy Processes in Finance: Pricing Financial Derivatives.* Wiley.

Schweizer, M. (1991) Option hedging for semimartingales. *Stochastic Processes and their Applications* **37**, 339–363.

Schweizer, M. (1995) On the minimal martingale measure and the Föllmer-Schweizer decomposition. *Stochastic Analysis and Applications* **13**, 573–599.

Schweizer, M. (1996) Approximation pricing and the variance-optimal martingale measure. *Annals of Probability* **24**, 206–236.

Schweizer, M. (2001) A guided tour through quadratic hedging approaches. *In: E. Jouini, J. Cvitanic, M. Musiela (eds.), Option Pricing, Interest Rates and Risk Management,* Cambridge University Press, 538–574.

Shiryaev, A.N. (1999) *Essentials of Stochastic Finance: Facts, Models, Theory.* World Scientific.

Steiger, G. (2005) *The Optimal Martingale Measure for Investors with Exponential Utility Function.* PhD Dissertation, ETH Zürich.

Stricker, C. (2004) Indifference pricing with exponential utility. In: *Seminar on*

Stochastic Analysis, Random Fields and Applications IV (R. Dalang, M. Dozzi and F. Russo, eds.), Birkhäuser, 325–330.

Tehranchi, M. (2009). Symmetric martingales and symmetric smiles. *Stochastic Processes and Their Applications* **119**, 3785–3797.

Vostrikova, L., Yor, M. (2000) Some invariance properties of Ocone's martingales. *Séminaire de Probabilités* **XXXIV**, 417–431.

Yong, J., Zhou, X.Y. (1999) *Stochastic Controls: Hamiltonian Systems and HJB Equations.* Springer.

Yor, M. (1978). Sous-espaces denses dans L^1 ou H^1 et représentation des martingales. *Séminaire de Probabilités* **XII**, 265–309.

Yor, M. (1992) *Some Aspects of Brownian Motion, Part I: Some Special Functionals.* Lectures in Mathematics. ETH Zürich. Birkhäuser.

Index